Telco Global Connect

TELCO GLOBAL CONNECT
Vol : 1

Author : Sadiq Malik

Copyright © 2015

ISBN-13: 978-1512099553

DEDICATION

For Dr Jean Grey ..my Phoenix Rising

This book is dedicated to the continued viability of the $1+ trillion Telco industry as data becomes the norm. It consolidates a series of informative articles that provide strategic insights for savvy Telco and Internet professionals on how to transform in the Digital era.

"Men are born soft and supple; dead they are stiff and hard. Plants are born tender and pliant; dead, they are brittle and dry. Thus whoever is stiff and inflexible is a disciple of death. Whoever is soft and yielding is a disciple of life. The hard and stiff will be broken. The soft and supple will prevail." — **Lao Tzu**

PREFACE

Telco professionals have already realised that unless they shift to new business models based on a refined customer understanding of customer expectations they will lose even more revenue to OTT players.

The transformation required by Telcos to capture new business opportunities necessitates building on existing capabilities and carefully stitching together strategies that unlock value in the convergence battlefield.

To become fit for the future, Telcos should combine growth and efficiency and effectiveness strategically by implementing a revolutionary lean and agile modus operandi. They will need to create and participate in new business and partnership models that will transform the global industry.

Telcos who do not wish to be relegated to the utility dumb pipe role must focus on developing new services, and compressing time to market for new services that customers will enjoy and gladly pay for. Their ability to learn how to monetize data will make or break their business models.

TABLE OF CONTENTS

Each chapter addresses a particular theme and the articles therein can be ready in any sequence.

Chapter 1 : Mobile World Congress Barcelona

This Chapter includes various posts compiled during the Author's visits to the MWC Congress in Barcelona. The material illustrates some of the highlights at the Congress and provide insights into the strategies of Global Telcos.The annual event provides the planet's best venue for mobile industry networking, new business opportunities and deal-making.

The Congress also hosts an exhibition with more than 2,000 companies displaying the cutting-edge products and technologies that define the future of mobile; the world's best opportunity for mobile industry networking; and the annual Global Mobile Awards ceremony, which recognises the most innovative mobile solutions and initiatives from around the world.

The mobile industry continues to be characterised by high levels of growth and opportunity, and as the industry becomes more dynamic, the opportunities within it increase in equal amount, as do the challenges. The Mobile World Congress conference programme provides an essential, in-depth coverage of the contemporary and future mobile industry, highlighting specific areas of growth and opportunity.

Chapter 2 : Key elements of Telco Transformation

This chapter highlights from a series of articles on the critical importance of the BSS/OSS in the delivery of new customer services while ensuring quality of customer experience and data monetization. CSPs have realized major opportunities in transformation that enables them to use their OSS/BSS investments more efficiently and even for totally new purposes.

With ARPU in decline, operators need new to be more adaptable and able to introduce new sources of revenue quickly. By adopting such a consolidated architecture for OSS/BSS, operators will be able to maintain control over costs while implementing network changes effectively.

Big Data technologies, and in particular their analytics abilities, offer a multitude of benefits to telecom companies including improved subscriber experience, building and maintaining smarter networks, reducing churn, and generation of new revenue streams.

Chapter 3 : The road to 4G monetization

This chapter delves into the various factors that contribute to the monetization of data as the Telcos forge ahead with 4G and Fttx in their access portfolios. To realize 4G's full potential and begin seeing a return on their extensive investments, service providers need back office systems that are just as robust, fast, flexible, and smart as the network.

LTE-enabled services are driving new revenues for operators but also pushing them to meet the increasing data demands of customers signing up for high speed mobile broadband offerings. Telcos are being impelled to deploy LTE as quickly but the the same time refresh their traditional operational support systems and business support systems while simultaneously launching innovative products that can justify their heavy investments in LTE.

The primary facets of a 4G monetization strategy involve speed to market, real time, flexibility of offer and charging/policy convergence, customer self-service over other interaction channels and the need for real time marketing analytics. To enable richer more compelling personalised services Telcos must deploy networks with capabilities, such as mobility, messaging, location, presence, profile and call control, and combine these with internet-style services such as social networking, search, advertising, direct marketing and mapping,

Chapter 4 : Network in the Cloud

This chapter exposes the various technologies that are propelling Telcos into the Cloud. From services to traditional hardware elements everything is becoming virtualized as Telcos start building out massive data centers.

The evolution of the internet has enabled the evolution of a new kind of business model – one built on virtual networks and software, available remotely for access by users no longer constrained by physical location. This in turn has resulted in the growth of a phenomenon known popularly as "Cloud Computing" or, a more encompassing term – "Services in the Cloud".

The phrase "Cloud" for these services arose due to their intangibility and virtual nature. This very virtualisation gives companies the opportunity to outsource and save valuable resources in terms of human and physical capital. The biggest benefit with the right service providers of Cloud Services is that enterprises get access to state-of-the art reliable technological solutions for their needs in an efficient manner.

Confidence in cloud services is still quite low and acts as a barrier to uptake. Here, operators play an important role in enhancing trust by offering secure solutions and thereby spurring development. Cloud service users naturally prefer buying from a known brand with good references, which gives the local telecommunications operator an edge.

Chapter 5 : Craft the winning business model

This chapter explores the arduous task of creating fresh and viable business models in the hyper competitive Internet era. All mature markets are becoming more mature and emerging market growth will slow in the next few years. New business models are required where operators make money from new customers and ecosystem partners rather than exclusively from end-users.

5.1	White Space : Digital Ecosystems in the New Mobile Web	148
5.2	Scenario Planning : Alternative futures for Telcos !!	153
5.3	Strategic Insight : Blue Ocean path for Telcos !!	157
5.4	SingTel : Red Dragon of the Telco World	163

How can companies create breakthroughs in value and performance? Most companies focus on matching and beating their rivals. As a result, their strategies tend to take on similar dimensions. What ensues is head-to-head competition based largely on incremental improvements in cost, quality, or both. So how do you leave rivals behind while sustaining spectacular growth for your company? Invent entirely new markets where no competitor has yet ventured. Asymmetric Business Models are among the most powerful economic tools that digital-age companies have at their disposal.

Telecom innovation like M2M, 4G, Ethernet and cloud computing is triggering the emergence of new business models, revenue streams and powerful efficiencies in several connected, adjacent industries Telcos will fail to capture revenue from fresh opportunities in Cloud Computing , M2M , and enterprise markets. In sectors such as healthcare , utilities and automotive Telcos can provide solutions that increase process efficiency and reduce transactional costs.

Chapter 1 : Mobile World Congress Barcelona

1.1 : MWC 14 Barcelona Wrap up : Quotable Quotes !!

 Finally the MWC14 is finished and all 85000 of us participants are back home , tired but reflective about the Congress. All kudos to the GSMA for organising this mammoth event with flawless precision : a Herculean labour indeed . And the city of Barcelona for being such gracious smiling hosts.

For me event was all about a brave new world forming , morphing. Its about 4G LTE connectivity becoming the Gold standard which will render cars into the "the new device". Algorithms analyze in real time and enable the intelligence infrastructure to make autonomous decisions.

Everything is going to be in the cloud as much as it is about the Internet of things (IoT) . Beeping sensors and software connecting humanity into a living breathing Web of real and virtual worlds. It will be about those who will create the future separating the winners and losers in the mobile industry.

SO HERE IS MY CHOICE OF THE BEST OF THE BEST quotes that captures the spirit and content of the Congress !!

"Everything that can be digitized will be digitized. Everything that can be connected will be connected. Bridging the infrastructure and ecosystem to digitization is a mega trend we see in the second phase of the Internet " (Tim Hottages , CEO Deutche Telecom)

"The industry focus in the coming years will be on personal data, connected living, digital commerce and networks. To realise the necessary developments in these areas, "huge investments of $1.7 trillion need to be done from now until 2020", (Jon Fredrik Baksaas,CEO of Telenor and chairman of the GSMA)

"Everybody worries about being a dumb pipe, and whether revenues will be able to support network investment that we need to make.Any other industry would be excited and highly optimistic given the strong demand in growth for their core services. However, the big problem we have as an industry is we have been unable to monetise this increased demand " (Chua Sock Koong , group CEO at SingTel)

"Communication networks are facing a lack of scalable and sustainable architecture to meet the challenges ahead in terms of data traffic increases, video uploads and downloads, and enhanced M2M communication. The network of the future has to be highly elastic in order to facilitate the adding or dropping of capacity and real-time provisioning of service. It needs to be highly orchestrated by key business imperatives, such as customer satisfaction, and it must be highly integrated so that synergies are fully embedded and captured across fixed and mobile, across borders and across segments."(Bruno Jacobfeuerborn Deutsche Telekom CTO)

"We have a very precise directive to establish asymmetric regulation where a company holds more than 50 per cent market share.This concentration of telecoms services to a few operators has resulted in uncompetitive pricing and obstructed new entrants. 70 million Mexicans have no access to broadband, while those that are connected experience low data speeds and poor quality. We need to leapfrog this situation." (Mexico Regulatory commissioner Adriana Labardini)

"Why are the next two billion not on the internet?. The reason is not because they don't have any money, it's because they don't know the value of having a data plan or the services they can access." (Mark Zuckerberg, founder and CEO of Facebook)

Regions around the world have always had initiatives regarding new technology, and it's important that you start early with R&D to stay in the forefront. For example, we first commenced work on 4G/LTE in 2000, and understand the need to invest in the research of 5G." (Ericsson's CEO Hans Vestberg)

"It's a market. One billion people in the world are disabled in some form or another, " (Mike Short, VP of Telefonica Europe)

"To change the vicious circle to a virtuous one, we're going to need more cooperation than there has been in the past. Then we're probably going to get to a win, win, win solution, as opposed to when you used to have people duking it out [as to who 'owns' the customer] in the past.There's greater value in cooperating than just competing. Developers cando this too by opening up their APIs"(Doug Webster, CMO, service provider, for Cisco)

"While big data is great, there is a lot of it out there in silos, and each data [set] speaks its own language, you need to be able to solve issues around homogenising it, then you need to solve issues around analysing the data... what we need is a 'universal translator.We also need a better policy over who owns this data.

We need to have joined up thinking about these open platforms,"
(Young Sohn, Samsung Electronics, president and CSO)

" NFV and SDN concepts are at the core of our strategy. These help us realise our future network vision, which is a mutli-service, multi-tenant platform where we can respond more quickly and efficiently to our customers' needs. With NFV, we're able to dynamically reroute traffic and add capacity without adding new boxes. With SDN, we're removing pre-defined physical limits of the network by shifting control from hardware to software. These allow the network to become simpler, more scalable. They also allow us to reduce costs significantly and more quickly address customer needs "(John Donovan, Senior VP– Technology and Network Operations, AT&T)

"Not everything can be automated at once and municipalities need to select carefully the value added services that will give consumers the most value for their initial offerings. Once a base is established the municipality can market the platform as an opportunity for other businesses that wish to deliver value added services. " Jeff Edlund, CTO, CMS, Enterprise Services, HP)

"Between 15 per cent and 20 per cent of subscribers say they have poor coverage at home, and yet more than half of homes in countries like the USA and the UK have Wi-Fi. Wi-Fi is an obvious choice for solving the problem." (Ken Kolderup, Kineto's CMO)

"We're attracting new vendors by offering them an alternative to a mobile ecosystem that depends on two restricted operating systems with many strings attached. Apple controls all aspects of their offering and eschews customization. Even on Android, from an OEM perspective, only one manufacturer is profiting. Many people in the industry want to see a third option become more viable, and our flexibility makes us an ideal partner." (Jay Sullivan, COO Mozilla)

" Mobile connectivity hold huge potential for women – currently an untapped economic potential. The Internet is proving to be an effective catalyst in transforming gender opinions. Access to a communication network provides women with the flexibility of working both remotely and on their own terms, allowing them to build independent companies. The more we can work together to develop this offering, the greater opportunities we can build for the region's women as a whole" (Dr Nasser Marafih, Group CEO, Ooredoo)

" The impact of MNP in developing markets is linked to two factors: the time taken to port numbers and the fee charged to the subscriber to use the facility. The porting time after submission of request varies from as long as two weeks in some countries to just a few minutes in others. In Ghana, for example, 92% of porting requests are completed within 5 minutes. By contrast, in Kenya, it is cheaper to buy a new SIM than to port from one operator to another. The Kenyan regulator attributes low uptake of the service to "the lowering tariff differentials between operators and the convenience brought about by dual SIM card mobile handsets. (Akanksha Sharma, Analyst, GSMA Intelligence)

"Our customers trust us, so our responsibility of how we manage this private information will become higher. Operators will have to become involved in defining how this sort of data is used, while protecting the privacy of the individual." (Kaoru Kato, Docomo's president and CEO)

"Operators are] shrinking in relevance too fast. Many partnerships (with content and internet players) have been for loyalty, not for revenue streams. We need to find different ways of charging our customers rather than complain " (Johan Dennelind TeliaSonera CEO)

"We will look back on this time and look at data as a natural resource that powered the 21st century," (IBM CEO Ginni Rometty)

Now let me tell you the best stop of my visit to MWC14 :)

El Celler de Can Roca in Girona (near Barcelona) :This restaurant has propelled Catalan cuisine on the World's Best Restaurants list. Dinners start off with caramelized olives, which come to the table on bonsai trees; after that, candied olives stuffed with anchovy are served on a miniature olive tree, then crispy shrimp followed by zucchini omelette and vermouth candy, then summer-truffle bonbon and brioche. Finish these nibbles, and we were ready for the actual meal ...yumm...need i say more... ps: the waiting list for this place is one year so you better book now to coincide with MWC15E

--♠--

1.2: MWC Barcelona Day 3 : The Connected Car is getting ready to roll

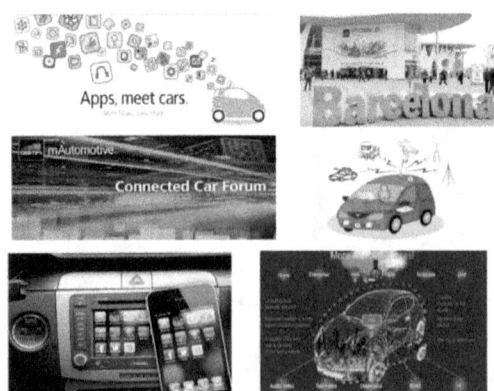

Browsing around the GSMA Connected Living pavilion is always inspiring because it is what mobile will do for our future.The Connected Car section exhibits exciting developments in the auto / mobile industries

partnerships. After all we love our cars and spend so much of our life commuting.For the neophytes a Connected Car is a car equipped with on board localization and communication technologies,internet access, and usually also with a wireless local area network.This allows the car to share internet access to other devices both inside as outside the vehicle and to interact with other vehicles and infrastructures.

Examples of Connected Car technologies are: In-vehicle navigation system with GPS and TMC for providing up-to-date traffic information, Adaptive cruise control (ACC), Lane departure warning system,Lane change assistance system, Collision avoidance system (Precrash system),Intelligent speed adaptation or intelligent speed advice (ISA) system, Night Vision system, Adaptive light control system, Automatic parking system,Traffic sign recognition system,Blind spot detection system, Driver drowsiness detection system,Vehicular communication systems,Electric vehicle warning sounds used in hybrids and plug-in electric vehicles, etc.

The opportunity for the Connected Car market is huge, both in terms of revenue and benefits, such as customer loyalty.The global connected car market will be worth €39 billion in 2018 up from €13 billion in 2012, according to new market forecasts.The market is close to the tipping point where connectivity in cars will become a mass market.As 4G LTE networks reach people , virtually every auto manufacturer is working toward a connected car that takes advantage of next-gen data speeds, from voice-controlled apps and infotainment to advanced diagnostics.

To build this new market, the mobile and the automotive industry will need to work in collaboration to surmount the challenges and deliver its promise.The industry is gearing up for a significant shift that will leave the landscape changed forever : whether you are a mobile network operator, automaker, software developer or hardware vendor there are huge opportunities on offer. In

2011,Machina Research conducted a market sizing study (M2M) Communication in the Automotive Sector, which forecast major growth in both telematics and infotainment services between 2010 and 2020.

Automakers indicated that:Telematics and infotainment will be offered across vehicle brands, with a critical mass on embedded solutions :Tethered solutions will continue, with a focus on providing upgradable solutions for technology and, hence, the higher bandwidth services, i.e. infotainment, high bandwidth apps (music & video) : Embedded solutions will continue for vehicle-centric,high-reliability and high availability apps (such as eCall and bCall) :Infotainment and video services are expected to grow exponentially.Machina's forecast for global wireless traffic generated by embedded mobility in the automotive sector shows entertainment and internet access driving an exponential increase in data traffic.

Mobile operators are seeking to better understand the auto industry's requirements with respect to: How in-vehicle services, and their connectivity requirements ,are evolving.How to enable all appropriate connectivity options for services. Greater understanding of these two aspects will facilitate the development of tailored approaches and services to support telematics and infotainment,in line with the underlying needs of automakers.

Moreover,cross industry collaboration will be required to overcome some existing ecosystem barriers. Mobile operators are particularly interested in fostering this joint collaboration in areas such as: Operational improvements on how to optimise data usage, common requirements for services and improving service delivery for different types of connectivity:new means to foster telematics and infotainment business development, such as through joint application programming interfaces (APIs), apps development and location-based services.

GSMA,along with a group of leading mobile operators,has already finalised the market requirements for the development of standardised embedded SIMs and for the remote management of SIMs.This has paved the way for the implementation of a worldwide-embedded SIM standard, reducing fragmentation and driving scale for 'connected' devices across various industries, including automotive, consumer electronics, healthcare and utilities.The goal of this initiative is to enable remote SIM management,helping drive global momentum for new, innovative and cost effective connected devices that will enhance daily life, while retaining the security and flexibility of current SIM card form factors.

AT&T was the first to enter the market with a proprietary, single global SIM platform giving automotive, consumer and M2M equipment makers the ability to work through a single carrier to wirelessly enable and connect products.Announced last year, AT&T's single SIM platform delivers built-in access to wireless and data networks throughout most of the world, with service in more than 200 countries and access to more than 600 carriers worldwide.AT&T spearheads two major initiatives to lead innovation in the connected car market – a first-of-its-kind connected car center in Atlanta, called the AT&T Drive Studio, and a modular, global automotive platform called AT&T Drive.

The AT&T Drive Studio integrates AT&T solutions across multiple companies and serves as a hub where AT&T can respond to needs of automotive manufacturers and the auto ecosystem at large.AT&T Drive Studio showcases end-to-end solutions that AT&T and its contributors can provide automotive manufacturers around the world.Significant ecosystem players are committed to the Drive Studio and will work alongside AT&T, including

Accenture, Amdocs, Clear Channel's iHeart Radio, Ericsson, Jasper Wireless, Synchronoss and VoiceBox.

A number of options exist to connect a vehicle, including: Embedded: Both the connectivity (modem and UICC) and intelligence is built directly into the vehicle Tethered: Connectivity is provided through external modems (via wired, Bluetooth or WiFi connections and/or UICCs), while the intelligence remains embedded in the vehicle.

Integrated: Connectivity is based upon integration between the vehicle and the owner's handset, in which all communication modules, UICC, and intelligence remains strictly on the phone. The human machine interface (HMI) generally remains in the vehicle (but not always). Each of these different connectivity options relies upon different mechanisms for linking the car to cellular technology.

As the M2M market grows, so does the maturity and intelligence of M2M cloud platforms, enabling intelligent devices, back-end systems and cloud platforms to seamlessly integrate. A M2M cloud platform could bring about a global solution for managing connected devices across different networks and interfaces. This is attractive to the end-user facing brand (typically the automaker) as it enables the performance of the device on the network to be visible and troubleshooting processes to be performed.

A key factor driving the Connected Car is that connectivity will be necessitated by regulatory mandates such as the European Commission initiative eCall, which calls for a system to be fitted to all new vehicles by 2015, meaning emergency services will automatically be contacted and given the vehicle location in the event of a serious accident. Automotive OEM manufacturers across the board are fully prepared for the eCall legislation, although to what extent and when it will finally be implemented is unclear.There is no doubt throughout the industry that the connected vehicle will provide significant advantages in terms of life saving solutions and stolen vehicle tracking.

So what do we envisage for the Connected Car over the next ten years? Futurist Ian Pearson (@timeguide) see the vehicle developing into a fully personalised, virtual environment with intelligent automation, creating a totally new relationship between the vehicle, the driver, and the passenger. As you get in, the seat will automatically move to your preferred position, as instructed by your phone. Even fabrics and other interior surfaces will be able to adapt their appearance and textures electronically to your taste. Heads-up displays let drivers keep their eyes on the road.

Many people will wear video visors that overlay data onto the field of view, making augmented reality a part of everyday life, and changing the appearance of everything around us, including car interiors and the world outside.In-car sensors will recognise and highlight points of interest and dangers ahead. Passengers will see an electronically enhanced world too, with information overlaid into their view, in addition to games, video entertainment and web access. Some of this data will come from the car and some from apps on their phone.

Further in the future, cars will come to you. They will take you where you want, and then you can just abandon them to go off to serve someone else. They will in effect offer a comfortable and socially inclusive form of public transport. This could even lead to buses disappearing from our streets WOW..exciting times ahead as the mobile and automotive industries work hard to connect cars and networks for our pleasure and their profit !!

---♠---

1.3: MWC 14 Barcelona Day 4 : Thinking Carbon and Diesel

I bet you did not know that the 2014 Mobile World Congress was the world's largest tradeshow certified to have a zero carbon footprint. The GSMA will certify the Congress as carbon neutral through the internationally recognised PAS 2060 standard. For several years, the GSMA has had climate change initiatives that have focused heavily on reducing waste in printed materials, encouraging the re-use and recycling of materials at the venue, utilising digital signage and electronic tools and working with Fira Barcelona, exhibitors and local partners to minimise the carbon footprint of the event.

To achieve carbon neutrality, the GSMA undertook activities to reduce the carbon footprint and will then purchase carbon credits to offset any remaining emissions. Carbon credits purchased by the GSMA will fund several CER (Certified Emission Reductions) projects certified by the United Nations Framework Convention on Climate Change (UNFCCC), including a hydropower project in China and a wind energy project in India, among others.

So what has this to do with us clever Telco people ?? Plenty.... unless you live under a rock you would know that telecommunications is supposed to reduce the need for transportation and the movement of people, as such total energy consumption should decrease in spite of the increased energy consumption needed for telecoms. Telco energy efficiency is such a critical issue that the GSMA Green Power for Mobile (GPM) programme, includes several initiatives such as awareness creation about the renewable technologies for telecom applications, CAPEX and OPEX analysis, vendor mapping and renewable energy market sizing.

The goal of GPM is to assist the mobile industry in adopting renewable energy sources, such as solar, wind, biomass, fuel cell or sustainable biofuels and hybrid power systems, in order to power an estimated 118,000 new or existing off-grid base station sites in the developing regions of the world. Reaching this target will reduce an estimated 2.5 billion litres of diesel consumption per annum and up to 6.8 million tonnes carbon emissions annually.

Due to an unreliable electrical power grid, Telcos / Tower Cos in Africa , India and other emerging markets use diesel generators, batteries and a variety of power management equipment to back-up the grid and ensure network availability. The growing cost of energy due to increasing diesel prices and concerns over rising greenhouse emissions have caused Telcos and TowerCos to focus on better power management methods.

Mobile network operators (MNOs) spend approximately US$15 billion on their annual energy use. Therefore, it is no surprise that energy efficiency is becoming a strategic priority for them globally. As mobile use continues to grow,so does the demand for energy, particularly by the network infrastructure.Lets take a quick look at the real cost of Diesel.A typical generator at a telecom site consumes 2.5 litres of diesel per run-hour. Add servicing ($0.76 per run-hour) and replacement ($1.05 per run-hour) costs to the fuel itself ($1.10 per litre) and you get a fully loaded cost of $4.56 per run-hour.

In Nigeria there are currently about 24,500 operational base station sites:12,000 are connected to the grid, of which approximately 80% need generator backup for regular grid outages lasting anything up to six hours per day plus 7,000 are generator-powered 24/7.And of the remaining 5,500, the vast majority are diesel-battery hybrids, with just a handful of systems also using renewable energy.

For the purposes of our calculation, let's assume that on average they run their generators for 12 hours per day. Pull that all together and diesel-related costs for existing sites add up to a staggering US$485M per year. And when you consider the country needs to increase the number of base stations to 60,000 by 2018, it's putting increasing pressure on operators whose power-related operating costs are skyrocketing while their subscribers are simultaneously pushing for lower prices. And this is just for Nigeria – one country (all be it the most populated one) in a continent of 55.

Renewable Energy Technology (RET) solutions like solar photovoltaic, wind power, biomass and fuel cells are the technologies of choice for alternative solutions at telecom towers today. Hybrid solutions that combine diesel generators with RETs and batteries are being customised. Fuel cells are being installed as a standalone solution replacing the existing diesel generator. In a limited number of cases where electrical grid availability is close to 20 hours a day or more, the diesel generator at the tower site has been replaced completely by enhancing the existing battery capacity leading to improvement in economics and reduction of carbon emissions on site. Batteries are and will continue to be a key part of any backup power solution.

India is one country that has played a pioneering role in the field of energy efficiency. Some of the initiatives that have been implemented in India so far include passive infrastructure sharing, replacement of old base transceiver stations (BTS) with new generation BTS, usage of outdoor BTS, optimised cooling at shelter, usage of intelligent transceivers (TRXs), reduction of air conditioner load by using cold ambient air for shelter cooling and operating air conditioners using stored energy in the batteries to reduce diesel consumption and carbon emission are. In the last four years with the evolution of technology, the typical power consumption of BTS has dropped by about 60% .

Bharti Infratel claim their introduction of Free Cooling Units (FCU) used in place of air conditioners has contributed to reduction of 4.1 million litres of diesel usage annually after deployment across 6,318 of its 34,220 tower sites. Technologies like solar photovoltaic, wind power, fuel cell and other renewable or clean energy sources have been deployed in about 4,021 telecom sites in India. Approximately 1,000 Indus Towers sites use solar photovoltaic to augment the grid and diesel generated power.Lets take another example : Nigeria and Ghana combined have a total of 30,000 + sites of which 50% are located in areas without commercial grid power , and mostly in rural and remote locations often with difficult accessibility for smooth and effective operations.

Therefore , operation and maintenance of the network remains a big challenge affecting the cost of operations network availability and reliability of mobile telecommunication services. Network OPEX will depend on various operational factors including energy supply, equipment maintenance, operational efficiency and robustness. The right technology, robust systems, right supply chain integration coupled with regular monitoring and reporting will enable the MNO to achieve OPEX efficiency and improve profitability. Operations and monitoring of deployed green power solutions is the most crucial part for guaranteed savings and expected performance.

In addition the MNO's must embed robust operational practices and a monitoring framework in order to address the challenges and mitigate the risks of theft, vandalism, and ensure site security. Site security is a major issue as there have been several cases of damage to tower assets across the region. This risk has hindered MNOs from investing in green power alternatives for powering the network. Thefts of equipment and fuel pilferage have affected the OPEX of telecom sites.

Meanwhile among the most developed nations the NTT group in Japan has worked hard on energy conservation, prompting them to introduce new sources of energy in the telecom field as part of the "Save Power" campaign from 1987. A total of 282 solar power systems generating a total power of 4,745 kW and 18 wind power generators producing a total power of 781 kW were introduced. NTT Facilities, which has taken responsibility for energy system design, architectural design, and building and energy management, proposed the new concept of "Green integration." This concept involves global environment protection and is based on NTT Group's past experience and expertise.

NTT Facilities has Green data centers, are operated in an environmentally friendly way and construction of the facilities are based on the "Green integration" concept. NTT Facilities has developed a FMACS airconditioning system, by taking into account the indoor air current, that can effectively cool such equipment, and the system eliminates high temperature areas and reduces the amount of energy consumed for air conditioning.

NTT also architected and deployed a MICRO GRID which combines various distributed power such as fuel cells, solar cells, and NaS batteries. The energy control system operates the distributed generators to control the influence on the commercial electric power lines wherein the micro grid is connected. This control system also optimizes the generation scheduling in terms reducing cost and environmental impact.

So what's the point of all this ?? GO GREEN OR GO HOME...unless you are fond of having 35% + of your monthly OPEX going up in diesel fumes !!

--♠--

1.4: MWC 2014 Barcelona : Godzilla and King Kong get down on it !!

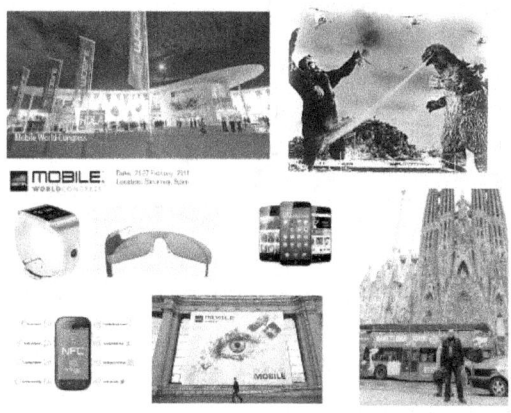

Most of you know that next week the Godzilla (Telcos) and King Kong (OTT) clans will congregate in Barcelona at the Mobile World Congress. This annual gathering of the leading (and not so leading) players in the Mobile and Internet value chain is becoming gargantuan. Those of you who are fortunate enough to attend , comme moi , should expect several topics being discussed and digested in the carnival atmosphere. Try not to get lost or bewildered in the 70000 strong horde of humanity !! At the end of 4 days of this Congress you will be tired but hopefully educated and inspired.

Plan carefully your visits to the exhibitor's , speakers and the GSM<A themed areas. The organisers expect 1,700 exhibitors utilising 94,000 net square metres of exhibition and business meeting space.And the Key Note presentations are not to be missed under any circumstances. Monday's opening keynote session sets the tone with a discussion about mobile operator strategies, which brings together chief executives from the US,

Europe, the Middle East and Asia. Regulators, too, are well represented. Over 80 government ministers are attending the government programme, which includes a focus on how to facilitate investment and bridge the catastrophic Digital Divide in third world countries.

So what can you expect in Barcelona next week ???

Most pundits agree that wearable devices will be a hot topic so expect lots of new products in this area. The industry is placing a lot of hope on this sector as the next big thing, but there's also an enormous amount of hype. Big questions still need to be answered. Do people actually want wearables? Which form factors work best? What will people pay for them? Will they keep using them? Right now the jury is most definitely out whether you will be sporting Google Glass to do something more useful than just showing off to your friends. The lack of true value that these devices add mean that it will be 2015 before they start to go mainstream. Maybe it is another category that Apple needs to redefine before we see real growth.

For smartphones it's all about iteration rather than innovation this year, although a few curved screen phones will offer something a bit different. We should see new devices from Samsung, HTC, Nokia, Huawei, TCL, Sony and others...as well as more devices with a curved/flexible display from Samsung and LG. However, flexible displays will only become popular when users can bend and fold a large screen into a small pocket. Other announcements are expected from Russian company Yota Devices, which could launch version 2 of its dual-screen Yota Phone. The possible launch of what has so far been dubbed the 'Blackphone', a joint project of Silent Circle and Spanish smartphone Geeksphone.

This year's MWC provides a good opportunity for Mozilla to announce the next step in Firefox OS development. It's an ideal time for an update since Firefox is also proving to be the best bet so far among the various operating systems that have emerged in the past year or two.Visitors to last year's event will remember that Mozilla created a storm by unveiling Firefox OS as an alternative HTML5-based operating system for low-cost Android smartphones in emerging markets.

The company said at MWC 2013 that more than 20 operators and handset vendors would support the new platform, and Spain's Telefonica has arguably been the staunchest supporter of the Firefox OS, launching the first device based on the OS in Spain last July. In November, Mozilla claims the OS was now available in 13 countries. MWC 2014 could prove to be a belated coming-out party for Tizen on Samsung devices. Tizen itself has also indicated that there could be an update on its operating system in Barcelona this year. I am really hoping to hear news on the commercial rollout of Mark Shuttleworth's UBUNTU OS for Mobile.

In the network, we'll see lots of SDN/NFV, small cells and LTE-Advanced. Thats means CLOUD CLOUD and more CLOUD. LTE's going to be a major focal point .There are 200 million LTE subscribers worldwide and CCS Insight forecasts they'll grow to over 1 billion by 2017. 5G probably is the hottest topic everyone likes to talk about, but only to say, "it's not defined yet and it won't be around for 5+ years." We love our acronyms in the Mobile Industry and we nod sagely to each other even if we don't have a clue on what the acronym means or how the technology works.

Most Operators will complain about the lack of spectrum to rollout national LTE networks as usual. Service-wise, we expect a lot about the future of voice/messaging and data monetisation (including OTT partnerships).But MNO's and OTT's (Godzilla and King Kong) will snipe at each other as to who is eating whose lunch (as usual). Who will win ?? Eventually both monsters will realize that they have to collaborate (rather than bash) with each other to provide valuable content and reliable connectivity at lower costs to the broadband consumer.

As if there is a shortage of people without high speed broadband on the planet !! Hey animals...who is going to pay for ugrading the Internet backbone as the existing infrastructure buckles under the enormous data traffic of high speed broadband networks and smart devices ?? Within its short lifespan, the Internet has already delivered huge economic and social benefits. For growth to continue, however, there must be a realignment of how investments are made and value is captured.

The real area of innovation in 2014 will be in the Internet of Things and specifically M2M. This category will see substantial growth across a number of verticals and we would expect to see a lot of companies demonstrating exciting solutions at the show. While there's obviously a huge opportunity here, there's still lots of work to do and it will be a long time before operators see any significant revenues from connected devices : GSMA Intelligence estimate that M2M made up just 2.8 per cent of global connections last year.As our lives become increasingly digital, the concept of identity in the digital world – and who authenticates s identities and guards their personal data – is going to be hugely important commercial opportunity for Telcos that move fast to capitalise.

In addition one would like to see the back office get the attention it deserves. To some extent, we've seen its profile rise in the face of hype around data analytics (aka, "big data"). Still, the OSS/BSS space isn't often considered exciting; that's a shame since it's at the center of carrier service innovation. The IT nerds have been working hard to add value to what was once the territory of the hardware engineers.

Embedded in millions of smartphones everywhere, unassuming NFC technology is now storming to the center stage. NFC is teaming up with Bluetooth and Wi-Fi to form a technology trio to power up the connected life vision . Together, these three technologies will transform mobile content sharing and entertainment by providing seamless hand-offs, instant and always-on connectivity, and new ways to work and game.

The latest generation of NFC tags can even switch on Bluetooth devices and Wi-Fi networks by harvesting their power, making it that much easier to access media and content. Instead of printing costly user manuals, electronics and games manufacturers can leverage the same NFC tags they're using for Bluetooth and Wi-Fi pairing to link consumers to graphic instructions and videos on their corporate website.

MWC continues to showcase the developed world benefits on an ongoing basis from incredible advancements in mobile technology and connectivity. Most Gurus agree that as an industry we need to do more to make mobile technology more accessible to a greater percentage of the world's population. The solutions that are being developed today have the opportunity to have wide scale economic, social and environmental benefits that will change people's lives.

YESS : VIVA LA TECNOLOGIA MOVIL : Bienvenidos a España....The City and people of Barcelona are great hosts !!

---♠--

1.5 : MWC 2015 Barcelona Day 2 : Most impressive Operator service launches !!

 The MWC is not short on various launches and pronouncements : smartphones , networking gear , software , middleware , alliances , operator initiatives etc. Under the cacophony of announcements the following 2 Operator launches caught my attention since they address very critical needs esp in developing countries : Education and Digital Commerce.

Malaysia's YTL combines internet and cloud to close digital divide !!

Malaysia's YTL, which has built out a nationwide 4G network to help close the country's digital divide, has also created what it says is the world's first national education cloud that is deployed in all 10,080 primary and secondary schools in coordination with the government's education transformation blueprint. With its 4G network, the company can offer anytime, anywhere learning. They have put in extra measures to create an VLAN-over-4G architecture to protect the children while they are connected. All it is backboned on a cloud-based learning platform, called Frog VLE (virtual learning environment), to create an intuitive experience for students, teachers and parents.

Launched in November 2010, YTL Communications was a pioneer in the delivery of mobile telephony on a mobile broadband network. Subscribers get both Internet and Voice in a single plan with voice roaming being free. Their 4G network is SIM-less and runs on a user ID that comes with its own mobile number. Subscribers can log on to multiple devices, all at the same time. Their Yes 4G service is a mobile internet platform that facilitates rapid and continuous innovation offering a ground breaking architecture that gives users unmatched performance, convenience and cost savings by bringing together mobile broadband, mobile telephony (voice and SMS) and cloud-based services across multiple devices, all in a single Yes account !!

YTL is the connectivity partner for the Malaysian Govt1BestariNet project which seeks to transform the education arena and establish Malaysia as a model of excellence in integrated, Internet-enabled learning. This to be achieved by providing all state schools with YTL 4G connectivity to Frog's Virtual Learning Environment (Frog VLE), which offers exceptional control enabling teachers, admin staff and even pupils to fully embed their learning platform into the school's working practices and tailor it to the needs of their school.

The Frog Virtual Learning Environment (VLE) is a web-based learning system that replicates real-world learning by integrating virtual equivalents of conventional concepts of education. For example, teachers can assign lessons, tests, and marks virtually, while students can submit homework and view their marks through the VLE. Parents can view school news and important documents while school administrators can organise their school calendars and disseminate school notices via the Internet.

Combining high-speed 4G internet access, a world class learning platform and access to 'best-in-class' resources and technology, with YTL , Malaysia is the first country in the world to bring its entire education community together on a single converged network designed specifically to meet the needs of teaching and learning !!

Telenor Digital adopts frictionless mobile login to fast track service uptake !!

Telenor Group is an international provider of tele, data and media communication services. Telenor Group has mobile operations in 13 markets in the Nordic region, Central and Eastern Europe and in Asia, as well as a voting stake of 42.95 per cent (economic stake 33 per cent) in VimpelCom Ltd., operating in 17 markets. Headquartered in Norway, Telenor Group is one of the world's major mobile operators with 176 mill mobile subscriptions per Q2 2014, revenues in 2014 of + NOK 107 billion and 33,000 employees worldwide .

Their Telenor Digital Unit has the challenging remit of developing services that will guide Telenor towards a future as an internet telco.Telenor Digital creates globally scalable solutions within next-generation communication services, cloud services, e-commerce, and the "Internet of Everything". Telenor Digital also enables global distribution of its own and third-party services and support new ventures within digital entrepreneurship.

Telenor believes that key to building a successful digital service is making it easy to use. Operators are ideally placed to securely remove obstacles to logging into an app or a web-page. By succeeding in the digital identity space, mobile operators will become a more visible part of consumers' everyday digital life. In response to this challenge the Telenor Global Backend is a common cloud¬-based infrastructure was developed to provide a global, shared system for giving Telenor's customers access to Internet service. This implies providing easy Sign Up and Log In – as well as frictionless payment.

As of today – nine out of 13 business units can offer payment of digital goods through Global Backend – e.g. on Google Play. In addition, six out of 13 Telenor BUs can offer services bundled together with a mobile subscription to their customers through the Global Backend infrastructure.

This also means that Telenor becomes more attractive towards partners because they have one integration point through the Global Backend, through which they can reach 172 million customers – instead of approaching 13 Business Units individually. At the heart of the global backend business model, is the customer ID – the unique key which all customer data is gathered around. The Connect ID is Telenor's global solution to authenticate end-users. In practice – it's a solution for signing up and logging into a service. In short, Connect ID offers easy access to all their services.

Across many industries, from entertainment to banking, from health to e-government, services and processes are becoming both more digital and more mobile, yielding efficiency and convenience benefits for individuals and businesses alike. However, consumers want to be able to access these services securely, shielded by robust privacy safeguards and strong data protection.

One of the GSMA's top four priority programmes, the Personal Data initiative has developed Mobile Connect, a fast and secure login system that enables individuals to access their online accounts with just a single click or, where appropriate, automatically. Mobile Connect can provide different levels of security, ranging from low-level website access to highly-secure bank-grade authentication. Mobile Connect promises to make passwords a thing of the past. To use the service, individuals subscribing to a participating operator simply need to click on a website's Mobile Connect button.

For the uninitiated the GSMA Mobile ID is a PKI-based secure authentication service that enables users of business applications to access secure accounts, platforms, applications and cloud services in a single, unified mechanism. The service both simplifies the user experience and protects the individual's identity as they interact in the digital world. Mobile ID is already preinstalled on the SIM card as a SIM toolkit (STK) applet, which can only be accessed by the mobile provider "over the air" (OTA) via the correct ID key. As a result, Mobile ID works on all mobile devices that meet the GSM standard (GSM 11.14 or 3GPP 31.111), regardless of the operating system.

For digital service providers such as Telenor, Mobile Connect will deliver the optimum balance between convenience, security and privacy. By enabling consumers to log in quickly and easily, it improves the customer experience and reduces the likelihood that authentication issues will lead to them abandoning transactions, while minimising fraud and errors. In this way, Mobile Connect increases customer confidence and strengthens loyalty, while enabling fast engagement and enhancing the digital service provider's brand.

1.6: Bienvenida MWC Barcelona 2015 : Separating the wheat from the chaff !!

With this year's theme of 'The Edge of Innovation', what can we expect from Mobile World Congress which has started in Barcelona today ?

The Internet of Things (IoT) has become shorthand for the concept of billions of connected devices. And it's an idea that's getting some real traction as big names position themselves as players in the market. FlexEnable has developed an organic transistor technology that that can be used to create truly flexible electronics with plastic flexible circuits. The company claims that this enables ultra-thin, ultralightweight, shatterproof electronic displays. This breakthrough technology, with cost-effective manufacturing, will revolutionise mobile devices and wearables and will open up new worlds for the Internet of things (IoT).

Carriers have been trying to figure out how to evolve their place in the IoT value chain beyond the mere connectivity piece, and people are excited about the potential of WebRTC in that respect. Following AT&T's Enhanced WebRTC API launch in January, So don't be surprised if there were more operator announcements along these lines at MWC, with Telefonica, Telenor and Orange likely candidates, as well as operators from the Asia Pacific region.

For the unitiated WebRTC (Web Real-Time Communication) is an API definition drafted by the World Wide Web Consortium (W3C) that supports browser-to-browser applications for voice calling, video chat, and P2P file sharing without the need of either internal or external plugins. WebRTC enables all kinds of real time communication such as audio, video and text between users by utilising the browsers.

This will be the year that 5G takes over MWC. Despite 5G technology not yet being standardised and unlikely to be ready for half a decade, many in the industry are ready to breathe hot air . While vendors jockey for mindshare, there will be no shortage of posturing around "what 5G is" and who should care. This includes the technologies that will make 5G a reality (including virtualisation, millimetre wave spectrum, unlicensed spectrum, duplex free operations) along with 5G drivers like ubiquitous IoT demands. Next Generation Mobile Networks Alliance feels that 5G should be rolled out by 2020 to meet business and consumer demands.

So what are the characteristics of a 5G network ?? For starters it is a super-efficient mobile network that delivers a better performing network for lower investment cost. It addresses the mobile network operators pressing need to see the unit cost of data transport falling at roughly the same rate as the volume of data demand is rising. It would be a leap forward in efficiency based on the IET Demand Attentive Network (DAN) philosophy.

5G is super-fast mobile network comprising the next generation of small cells densely clustered together to give a contiguous coverage over at least urban areas and gets the world to the final frontier for true "wide area mobility". It would require access to spectrum under 4 GHz perhaps via the world's first global implementation of Dynamic Spectrum Access. And finally it is a converged fiber-wireless network that uses, for the first time for wireless Internet access, the millimeter wave bands (20 – 60

GHz) so as to allow very wide bandwidth radio channels able to support data access speeds of up to 10 Gbit/s. The connection essentially comprises "short" wireless links on the end of local fiber optic cable. It would be more a "nomadic" service (like WiFi) rather than a wide area "mobile" service.

Middle East operator Etisalat said the rollout of 5G technology is one of its goals for 2020 and will underpin its future support for machine-to machine (M2M) and eGovernment services as well as the wider Internet of Things (IoT).

While both 3G and 4G radio access networks (RANs) were built as stand-alone network, 5G RAN will be deployed by integrating the existing LTE-Advanced (LTEA), its evolution technologies, and new radio access technologies (RATs). So expect Virtualization and SDN (the other big thing in network evolution) will be extended to 5G mobile wireless networks as well. As you know with wireless network virtualization, network infrastructure can be decoupled from the services that it provides, where differentiated services can coexist on the same infrastructure, maximizing its utilization.

Consequently, multiple wireless virtual networks operated by different service providers (SPs) can dynamically share the physical substrate wireless networks operated by mobile network operators (MNOs). Since wireless network virtualization enables the sharing of infrastructure and radio spectrum resources, the capital expenses (CapEx) and operation expenses (OpEx) of wireless (radio) access networks (RANs), as well as core networks (CNs), can be reduced significantly.

Many telecoms companies are now pursuing multi-play strategies in which they offer a number of services, including mobile and fixed voice services, broadband and pay-TV. Vodafone has been active in this market with several high profile acquisitions in Europe, while the deal between BT and EE in the UK will also create a quadplay powerhouse. It would therefore be a major surprise if this isn't a hot topic when executives from the major mobile operators assemble at MWC. CCS Insight research suggests that consumers are interested in signing up to companies offering this range of communication and media services if their offering is good value. The appetite for multiplay services may also be driven by people owning an increasing number of devices.

Behind the launch hoopla of super sophisticated smartphone (with their wretched battery life) look out for another technology that is gaining momentum. NFC ..aah yes Near Field Communications...thats what some congress participants will be using to register or pay for their cappucino !!! NFC is a contactless radio technology that can transmit data at speeds of up to 424kpbs between two devices within four centimetres of each other. Some plastic debit and credit cards now contain NFC chips, enabling people to pay for items by simply tapping the card against an NFC terminal. Mobile phones are also being equipped with NFC capabilities, enabling consumers to use the technology to interact with readers to access information, validate tickets, redeem vouchers, collect loyalty points, make in-store payments and use many other commercial services.

Derived from well-established radio frequency identification (RFID) technology, NFC builds on the existing standards used for contactless card payments. NFC-enabled devices are, therefore, generally compatible with existing contactless card terminals in retail outlets and restaurants. Across the world, businesses and consumers are looking to digital commerce to provide flexible and efficient transaction services across a range sectors, including retail, transport, financial services, online and advertising. As a result, the roll out of digital commerce services is gathering pace.

By the end of 2014, there were more than more than 150 SIM based NFC launches, of which nearly 60 operate as commercial services around the world.NFC has widespread support from device makers. Some analysts report a tsunami momentum and NFC smartphones are reaching a tipping point in 2013 with unit shipments set to surge 156% this year to 400 million, according to Strategy Analytics. At the same time, the number of NFC-ready point-of-sales terminals is set to rise dramatically from 6.7 million in 2012 to 44.6 million in 2017, according to Berg Insight.

With the release of Android 4.4, Google introduced a new platform support for secure NFC-based transactions through Host Card Emulation (HCE), for payments, loyalty programs, card access, transit passes, and other custom services. With HCE, any app on an Android 4.4 device can emulate an NFC smart card, letting users tap to initiate transactions with an app of their choice. Apps can also use a new Reader Mode so as to act as readers for HCE cards and other NFC-based transactions. On September 9, 2014, Apple also announced support for NFC-powered transactions as part of their Apple Pay program. Apple stated that their version of NFC payment is more secure than competitors because Apple Pay implements tokenization of its data in order to encrypt it and protect it from unauthorized usage. And we all know that the technology Apple chooses to embed in their Iphone hits mainstream. As the WiFi accionados.

Anyone who has attended Mobile World Congress (MWC) over the last two years will know that the GSMA Connected City has been the place to experience the very latest in cutting-edge mobile technology. This year, it's been rebranded the GSMA Innovation City, to reflect the great technological strides that the mobile industry continues to make, as well as the number of innovative, state-of-the-art products and services that will be on display.

The Innovation City is all about experiencing and exploring the latest mobile technology in the context of an extensive virtual cityscape, where amenities, services and businesses are transformed and given added functionality through mobile connectivity. At the City, you can see a demonstration of an

International Coupon System which allows purchases to be made across the globe with a smartphone, via a one-tap, pay-and-redeem digital wallet which is being run by the GSMA Digital Commerce programme.

Of particular interest in the M2M space will be two demos of live operator profile swaps based on the GSMA Embedded SIM Specification, which is intended to make it easier to remotely change the profile of SIMs that are often located in difficult to reach or hazardous locations. One demonstration will show a smart parking application using a profile swap, and the other will demonstrate a connected car profile swap from the perspectives of both the end consumer and the backend provider.

So learn and enjoy MWC 15 even as you navigate in the hustle and bustle of 100000 attendees !!

--♠--

1.7: MWC15 Barcelona Wrap Up : Technologies , Trends and Quotes

 MWC is like a small city within the city of Barcelona, so imagine the excitement of meandering through this conference with all its 90000+ attendees gawking and awing at the technologies that will shape the world in the next decade.It is estimated that the Event has

raked in some €450 million for the local economy .No wonder the entire City of Barcelona gears up to welcome the attendees.

There was of course no shortage of quotes , comments and pithy remarks from the Speakers across the Telco value chain. So for this summary i have grouped quotes under some technology themes that caught my attention .You can derive vital clues to calibrate your technology investment and business strategy plans from the quotes herebelow.

5G

" The mobile industry should consider 5G as a "special generation", introducing challenges in all layers of the technology" (Mike Short, VP of public affairs Telefónica Europe)

" We have engineering teams working on LTE and 5G. Each time the 5G team unveil a new performance leap, the LTE engineers respond by matching it " (Matt Grob, VP & CTO for Qualcomm)

"We really do need 5G in order to have a paradigm shift. The order of magnitude jump in traffic is what is really driving this move." (Mischa Dohler, professor of wireless communications at King's College London)

"5G is a fundamental change in technology and will have a significant impact on how we offer services. We must look at performance and coverage, and not just consider microcells." (Allan Kock, director of RAN development at TeliaSonera)

" 5G must have green as part of its very DNA, ensuring that all aspects of 5G, from access networks, data centres and transport network to connected devices only consume energy when they are being used. However 4G is a success, let's enjoy . We shouldn't jump too fast," " Stephane Richard Orange France

"Most of our competitors talk about 5Gb/s and 10Gb/s or some other number, but they're not telling you the configuration, which is ridiculous. 5G should not be a "forced leap" in technology, but draw heavily on re-use existing radio air interfaces – LTE, WiFi and LTE-U – under a common control plane." (Marcus Weldon, CTO of Alcatel-Lucent)

"Computing costs have fallen 1000-fold since their inception.We've achieved this change with semiconductors, we now have to do the same with 5G compared to 2G," KT Telecom CEO Chang-Gyu Hwang

Ultimately, the success of 5G will depend on the success of the entire ecosystem, one in which innovation will become the key driver behind 5G development market demand.In the GSMA hyper-connected vision mobile operators would create a blend of pre-existing technologies covering 2G, 3G, 4G, Wi-fi and others to allow higher coverage and availability, and higher network density in terms of cells and devices, with the key differentiator being greater connectivity as an enabler for Machine-to-Machine (M2M) services and the Internet of Things (IoT)..

Bottom Line on 5 G : Hold your breath till 2020 if you insist on 5G or focus on monetising LTE and LTA A now !!

Telco Vs OTT

" There is a clash of business models between network operators and internet firms, and that network players were left holding the short straw. We are an asset-heavy industry, where everything is interoperable and open while OTT is asset light. How can you compete with a voice, SMS or video service which costs nothing?" (Timotheus Hoettges, Deutsche Telekom Group CEO)

" Operators are competitively disadvantaged with over-the-top players, owing to heavy infrastructure investment. Regulators should apply the principle of "same service, same rules" to players of every hue, and that consumers should have a "portable digital life" where it's just as easy to switch digital ecosystems as it is networks. (Caesar Alierta , CEO Telefonica)

" Offering free basic internet services can help mobile operators grow their businesses faster in emerging markets. And once users get a free taste of the internet they'll be more inclined to pay for mobile data.The Ebola crisis in West Africa is an example of a place where a crisis may have been exacerbated by the lack of good connectivity," " (Mark Zuckerberg , CEO Facebook)

"Our goal here is to drive a set of innovations (like Project Loon). I think we are at the stage when it's important to think about hardware, software and connectivity together. We are very happy to work with Facebook on Internet.org" " (Sundar Pichai, SVP of products at Google)

"Those opposed to open internet rules, they like to say we used depression era regulation. But we took Title II and modernised it. We built our model on a regulatory model that has been wildly successful in the US for mobile." (Tom Wheeler FCC Chairman)

"I think you almost have two choices as an operator: either you say that you can be very, very efficient with pipes and that somebody else can deal with all of that and you're going to have the lowest cost and the best quality on my pipe, transmit as much data as possible " (Hélène Barnekow TeliaSonera CCO)

It all seems so unfair that OTT players make their money by loading more and more traffic on the operators' networks, thus causing the operator to pay for network capacity upgrades to facilitate the OTTs' increasing VoIP and IM traffic that was the root cause for decreases in the operators' core voice and SMS revenues. Telcos you can't wish the OTT's away anymore than you can do the same with your mother in law !!

Bottom Line : Collaborate or compete the consumer will have his cake and eat it too !!

Telco M+A

"We are clearly looking for distressed assets. If we can find the right opportunity, within specific countries, we will do that, "Big operators buying up whatever they can are rarely a success case. It bites them in the behind later. Operators "should start with in market consolidation, then fixed mobile convergence ... then international – the latter is very complicated" (Mats Granryd, CEO of Tele2)

"There is room for consolidation in Africa. Our focus is on building up our position in its existing markets " (Marc Rennard Orange)

"Authorities are starting to understand that the industry is setting itself up for a data only world. The Europen Commission is a bit of a Dr Jekyll and Mr Hyde. When you discuss the merits of a deal they are extremely conservative. When you come with convergence and OTT as an argument they are not very receptive. The commission views market consolidation as a question of] fixed cost efficiencies," (Thomas Wessely, partner at Freshfields)

" The value of mergers and acquisition in the technology and operator sector grew 34.9 per cent in 2014 from the previous year, according to. The last time it was that high was in 2000," said Thrasher. And last year telecom M+A as represented three of the ten biggest deals" (B. Holt Thrasher, MD for Mooreland Partners)

" We believe M+A as a potential means to transform its business as the industry moves towards software based service delivery. Our previous approach was more focused on filling gaps. We are now looking at a complete view of which new business we need to be in. The performance on acquisition is not necessarily stellar. It's an important caveat. An acquisition is not a panacea, it's not going to solve the problem" (Rima Qureshi, SVP, head of M+A, Ericsson)

There is no substitute for old-fashioned focus on the fundamentals of MA: a clearly articulated and well thought-out strategic rationale for the acquisition should become the yardstick by which to measure individual decisions that arise during the course of a transaction. Despite the clear economic rationale, there may still be some obstacles to consolidation, such as the unrealistic ambitions of owners and senior management or an investor reluctance to take a financial write down. While in-country MA has led to improved profit margins for the merged business, this has, in most cases, come at the price of a loss of market share

Bottom Line : Do your Due Diligence because mating two turkeys does not an eagle make !!

Internet of Things (IoT)

"The pace of change in today's mobile ecosystem sometimes makes it difficult to keep focus on core business areas like roaming that still carry tens of billions of dollars in revenue opportunity, Things (IoT) and wearables. It warned mobile operators that these newer innovations could be "market distractions" and carry an opportunity cost of up to a staggering $46 billion." (Mary Clark, Syniverse)

" Given that M2M ARPUs are low, the cost of supporting sensors has to be significantly below that of supporting subscribers. If operators are going to cost-effectively manage a huge number of new devices in their networks then they need an understanding of the machine's environment and enough information to trigger rapid automated responses to any changes that affect service performance (Anukool Lakhina, CEO and Founder of Guavus)

"The car of the future is a smartphone on wheels." Dieter Zetsche, Chair of Daimler AG

" We believe cyber security and approval from regulatory authorities will be key challenges when it comes to connected cars, but says both Nissan and Renault know exactly where they are headed with "autonomous" cars (not driverless). (Carlos Ghosn CEO at Renault – Nissan Alliance "

" It says something about the achievements of the mobile industry over recent years that many IoT propositions appear to take the mobile connectivity component for granted: there is an assumption that it will simply work. Interoperability will remain a critical success factor for IoT. Interoperability still matters. Any IoT device with poor interoperability will struggle to achieve scale. " (Lars Nielsen, General Manager, Global Certification Forum)

" From an ATT perspective, we are providing an end-to-end (E2E) platform that enables a unified experience across a wide range of devices and one-off capabilities in the market today. There's a lot of noise in the marketplace that can make it complicated for a consumer or a business to make sense of this notion of the Internet of Things. It is important for us to show real benefits, whether you're a consumer or business " (Kevin Petersen, ATT Digital Life President)

" The SIM card is evolving to support future new services in both the machine-to machine (M2M) and consumer markets. By 2020 we forecast nearly 1 billion cellular M2M connections and 9 billion consumer connections that will require SIM cards, so it is critical to ensure security and robustness in the evolution of the SIM." (Hyunmi Yang, chief strategy officer GSMA)

" In the emerging Internet of Things (IoT) space telecoms operators will likely continue to provide connections into consumers' homes and power the different screens. But to remain relevant within the connected home landscape , operators will need to find new ways to interact with the things people care most about in their homes – things that help them stay comfortable, help keep them safe and help them save energy. (Chris Borros Nest Labs)

"It's (IoT and Digital Life Platform) not just a differentiator, it's an imperative for success. Nearly every CIO I talk to has security as his or her number one concern. What IoT can do for businesses is so exciting, but customers want to know their data is secure," (Ralph de la Vega, president and CEO of ATT Mobile and Business solutions)

Wearable computers, personalized advertising billboards and self driving cars are some of the items that appeared in science fiction blockbusters like the Minority Report to depict a futuristic world. Now, in 2015, this scene has become a reality. With the spread of smart devices, evolution of mobile networks and growth of smart technologies encompassing sensors, cloud computing and Big Data, the era of the Internet of Things (IoT) has finally arrived.

Bottom Line : Jump on the IoT bandwagon for the right reasons. Be prepared.. its not an trial initiative but a way of life !!

Finally the GSMA has demonstrated that they can grow and handle an event of this magnitude with M2M efficiency. Besides the GSMA is a treasure trove of stats and analysis which is available to members and subscribers. The Event was more about exciting new technologies rather than existing Telco problems.

As such the best view point came from Jon Fredrik Baksaas, Telenor Group CEO, when he said " 70 per cent of the operator's customers switch off data roaming when travelling abroad. But in the future we want to get rid of bill shock, so that mobile systems basically serve us. They deliver better peer-to-peer connectivity, better security and better privacy in the long run. Removing the bill shock phenomenon needs new pricing initiatives to create a new level of affordability".

So while Mobile Operators invest billions in the mind boggling capabilities of new technologies that will they should also try to resolve their subscribers current pain points..like BILL SHOCK. That will earn you new revenues , improve CoE , reduce churn...and less cause to complain about the OTT's eating your lunch !!

---♠--

Chapter 2 : The roadmap to BSS/OSS and Transformation

2.1: BSS/OSS Transformation : Absolute Imperative in the 4G world

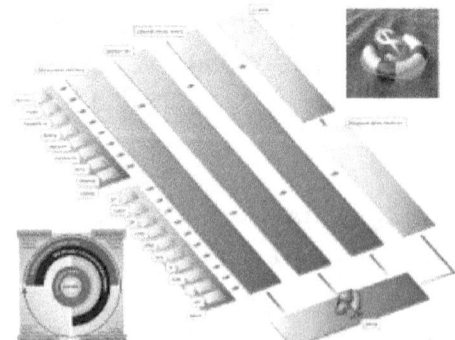 To remain competitive, Communications Service Providers (CSPs) must take a holistic view of how they will achieve the business and technical agility and flexibility to continue to compete effectively. The need to launch new types of services, based on new business models, on

multiple platforms and in multiple industries necessitates a need to review, renew and transform OSS/BSS to increase the flexibility and reduce time to market.

BSS systems (typically including billing and CRM), have always been separate from OSS systems (such as resource management, service activation, provisioning, fault management, etc.), which included having separate business processes and people. Over the years IT infrastructures have evolved into expensive, complex collections of monolithic applications interconnected with specially built point-to-point interfaces. A highly adaptable and flexible OSS architecture is also the key enabler for more streamlined operations.

Alongside the migration to a Service-oriented Architecture (SOA), the most significant new demands on OSS/BSS technologies come from the need to capture and integrate various sources of information as service providers introduce new blended services. Rather than tracking and fulfilling dedicated services in a siloed manner, OSS/BSS systems will need to take into consideration key operational and business variables as services are poured into a melting pot and served to customers in a converged package.

The transformation to the next generation OSS is an evolution not revolution because the current OSS systems are crucial for the operation of a network they cannot be replaced overnight. The transformation and migration will need to happen gradually, making the challenge even greater – old systems cannot be turned off before new systems are in place. A solid transformation strategy is based on SOA and ETOM principles to enable faster and more efficient service cycle times for new products and services.

EA is intended to bridge the gap between strategic planning and implementation efforts. To bridge that gap requires a process that is holistic : the process must cover the impact of strategic business change on technology, business processes and information. The recommended approach to enterprise architecture (EA) unifies three fundamental practices: business architecture, technology architecture and information architecture. EA is intended to broadly influence and support investment decisions and organizational change.A properly resourced and well-run EA program is essential to achieving and communicating the promised benefits.

The IT industry has embraced the concept of a Service-Oriented Architecture (SOA) as a standardized, more efficient way to build enterprise IT infrastructures. SOA, together with a revised enterprise business process, is the right way to build BSS and OSS applications, because it supports more agile internal operations, enables interoperability among new applications, and can be used to leverage existing BSS and OSS assets by adapting them to the SOA model.

The architectural vision is formulated, to act as a beacon guiding decisions during the rest of system structuring. Architecting the future state of EA is the heart of the entire process. The goal is to translate business strategy into a set of prescriptive guidance to be used by the organization (business and IT) in projects that implement change. Enterprises can engage in different types of strategic business visioning. One of the goals of business strategy is to strike a balance between long-term strategy (traditional strategic planning) and the strategies to be pursued as a result of a short-term opportunity

eTOM is part of the New Generation Operations Systems and Software (NGOSS) standard, developed by the Tele Management Forum (TM Forum). NGOSS is a comprehensive, integrated framework for developing, procuring, and deploying operations and business support systems (OSSs/BSSs) and software. It is available as a toolkit of industry-agreed specifications and guidelines that cover key business and technical areas including:

• eTOM Business Process Map: An industry-agreed set of integrated business process descriptions, created with today's customer-centric market in mind, used for mapping and analyzing operational processes.
• Shared Information/Data (SID) Model: Comprehensive, standardized information definitions acting as the common language for all data to be used in NGOSS-based applications. A common information language is the linchpin in creating easy-to-integrate software solutions.
• Technology Neutral Architecture: Key architectural guidelines and specifications to ensure high levels of flow through amongst diverse systems and components.
• Compliance and Conformance Criteria: Guidelines and tests to ensure that systems defined and developed utilizing NGOSS specifications will interoperate.
• Lifecycle and Methodology: Processes and artifacts that allow developers and integrators to use the toolset to develop NGOSS-based solutions employing a standard approach.

In 2008, Belgacom launched a Business Transformation programme with the objective to implement a new operational model for the group at the 2016 horizon. The expected business benefits are quantified via measurable indicators (KGI/KPI) mapped to the components of the model and are enabled by a major evolution of the BSS & OSS platforms. The TM Forum Solution Frameworks NGOSS framework, tuned to the Belgacom model and needs, helped Business & IT together aligning the B/OSS roadmaps to the benefits.

Transforming OSS/BSS platforms and partially rebuilding with more modern and harmonized platform components leads to considerable savings in the long run: reductions in time, effort and costs spent on system integration, administration, maintenance and training. An ideal vision of harmonized OSS architecture is inspired by the TMF Lean Operator Initiative and based on four key areas:

• CSP's process architecture: The CSPs' business processes can be supported by introducing modifiable operator process templates out-of-the-box and enabling a higher level of automation in their daily routines.

• Common information architecture: stepping away from "stove piped" data and supporting shared information and data models. This enables OSS/BSS level application interoperability through Common Information Models.

• Modular application architecture: will bridge the gap between service and resource management applications. A high level of modularity allows flexible solution building: it enables easier maintenance, allows changes on one component without affecting others and allows new components to be added as required.

• Application integration architecture: Interoperability and time to market is improved through compatible interfaces, common information models and through leveraging partner ecosystems and productized adaptation libraries.

There is no doubt that the major challenge for CSPs today is providing higher value for end users while facing constant, intense cost pressures and operating in often complex value networks. Transforming OSS/BSS platforms and partially rebuilding with more modern and harmonized platform components leads to considerable savings in the long run: reductions in time, effort and costs spent on system integration, administration, maintenance and training.

To keep leadership in the Brazilian market, NET had to improve its technology platform, transform its operational processes and change its field services operation. Challenges included the adoption of a field force management system by a distributed organization. The key success factors included: clear strategy/ business requirements , prioritizing the deployment of capacities, well defined governance structure , pragmatic planning and management methodology amd the choice of a field force management solution.

The new, digitally connected world is driving transformation, bringing with it new players, advanced applications, broadband services and higher QoS demands. Seizing this transformation by making your business and technology evolution work together is the key to profiting from market changes.BT has undergone a major transformation and continues to change. What was a traditional networks-focused, telco R&D organisation is now a 'softco', centred on developing software and using software development methodologies and practices. At the same time, networks and computing are quickly converging into what we know as cloud computing. Not surprisingly, BT's continuing transformation is now addressing the cloud services market.

According to the TM Forum, every dollar spent on traditional approaches to OSS application development results in as much as $5 USD (€3.37) of integration activities to address the impact on all related systems and networks. Properly implemented, next-generation OSS applications will reduce the integration costs. But even more beneficial, the more modern OSS applications will leverage commodity hardware and software and expand reuse of services. This will reduce both CAPEX and OPEX loads on the balance sheet.

2.2: Enterprise Architecture (EA) : The road to BSS/OSS Transformation

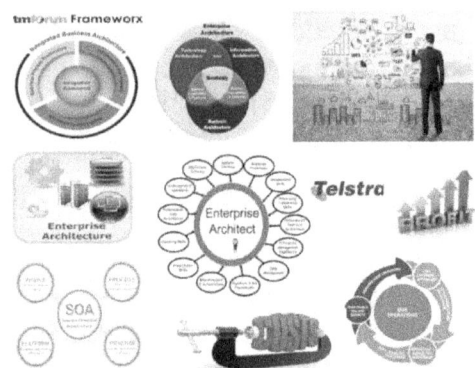

BSS systems (typically including billing and CRM), have always been separate from OSS systems (such as resource management, service activation, provisioning, fault management, etc.), which included having separate business processes and people. For example, revenue focused BSS was always run by the IT department, and cost-focused OSS was run by network operations. This traditional binary approach would have likely continued to be sufficient if not for the major transformation the telecommunications industry is undergoing, where service providers are becoming retailers of multimedia and entertainment services.

Over the years Telco IT infrastructures have evolved into expensive, complex collections of monolithic applications interconnected with specially built point-to-point interfaces. Transforming OSS/BSS platforms and partially rebuilding with more modern and harmonized platform components leads to considerable savings in the long run: reductions in time, effort and costs spent on system integration, administration, maintenance and training.

An enduring vision of harmonized OSS architecture is inspired by the TMF Lean Operator Initiative and based on four key areas:
• CSP's process architecture: The CSPs' business processes can

be supported by introducing modifiable operator process templates out-of-the-box and enabling a higher level of automation in their daily routines.

• Common information architecture: stepping away from "stove piped" data and supporting shared information and data models. This enables OSS/BSS level application interoperability through Common Information Models.

• Modular application architecture: will bridge the gap between service and resource management applications. A high level of modularity allows flexible solution building: it enables easier maintenance, allows changes on one component without affecting others and allows new components to be added as required.

• Application integration architecture: Interoperability and time to market is improved through compatible interfaces, common information models and through leveraging partner ecosystems and productized adaptation libraries.

Bear in mind that the transformation to the next generation OSS is a revolution, nor is it fixed to a particular date or year. As current OSS systems are crucial for the operation of a network they cannot be replaced overnight. The transformation and migration will need to happen gradually, making the challenge even greater – old systems cannot be turned off before new systems are in place. To mitigate these risks, future needs must be anticipated in advance and OSS architecture must be designed to fit with future requirements from the start. OPEX for the legacy OSS needs to be reduced to make room for new investments and replacement of the old functionality. OPEX reduction takes many forms, including:

• Removal of old, redundant OSS applications and systems
• Streamlining of functionality in legacy OSS

- Replacement of bespoke/ customized systems integration work with standards-based software and off-the shelf mediations
- Selective freezing of legacy OSS applications and systems
- Encapsulation of functionality and making it "service aware" with SOA
- Effective use of key OSS systems, moving functionality to these and taking other systems off-line

The Enterprise Architecture Model describes the elements of business – strategy, business cases, business models, processes, supporting technologies, policies, and infrastructures that make up an enterprise. It also provides means for governing the enterprise and its information systems, and planning changes to improve the integrity and flexibility. In other words, Enterprise Architecture crystallizes the organization – what it has to do and how – to be as efficient and productive as possible.

In the Enterprise Architecture, the business quadrant handles the value chain aspects relevant to the business as a whole: where to improve the business efficiency and develop new value propositions and how to increase efficiency and competitiveness of the business in the context of its environment: markets, competitors, legislative and environmental aspects, influences and impacts.

With Enterprise Architecture (EA), new opportunities and capabilities will raise some real competitive advantages for Telco operators. Architecting the future state of EA is the heart of the entire process. The goal is to translate business strategy into a set of prescriptive guidance to be used by the organization (business and IT) in projects that implement change. As such EA is a process discipline. Done well, it becomes an institutionalized part of how an organization makes decisions to direct its investments, such that the chosen business strategy will be realized. Usually a system is seen as a necessary cost to make the business – not anymore and certainly not with EA !!

Managing IT complexity to support business strategy is a big challenge for enterprise architects at large companies when a company has global operations, as is the case for Telstra, an Asia-based telecommunications firm. However Telstra's enterprise architecture (EA) team addressed its challenges by focusing on customer engagement, improved agility, and global business strategy enablement.The EA process bridges the gap that otherwise exists between business strategy and technology implementation. High-performing organizations are process-disciplined which is lacking in many of the Tier 2 Telco operators.

In turn, every high-performing process must be defined and documented, have process owners and be closed-loop with governance in place. Activities in this phase of the architecture process include but not limited to :
• Scoping the EA program and the next iteration thereof in terms of breadth and depth, which is known as defining what is meant by "enterprise"
• Gaining executive sponsorship and support
• Conducting stakeholder analysis
• Identifying the EA leader or chief architect
• Building and chartering the "EA team," which will own and facilitate the EA process and establishing clear roles and responsibilities
• Assessing organizational readiness and EA maturity
• Developing an initial communications plan, communicating the role of EA and setting expectations of individuals participating in the process
• Establishing a plan for setting up a governance mechanism
• Defining measures of success to articulate value delivered

Currently many Telcos are burdened by a wide range of systems 'isolated' for the operation of its business. This reality does not allow effective sharing of information between systems and / or applications. In recent years they have acquired several technologies were acquired from different manufacturers and suppliers, most could be considered islands of information and technologies. Today's service providers must close the gap between their Customer-facing BSS and network-facing OSS. With Enterprise Architecture (EA), new opportunities and capabilities will raise some real competitive advantages. As an example, let's consider a set of typical (separated) systems:

• Automation system: The principle behind this is to improve efficiency, automating several steps (or all steps) of certain tasks. Since the tasks are automatic, the delay is caused by latency of the system itself, leading to execution of thousands of tasks per second instead of seconds/minutes spent in each task.
• Customer segmentation: Groups people according to attributes that store information relevant for understanding customer behavior, and can be used to predict the probability of acceptance or refusal of a certain product or probable churn.
• CRM tool: Contains all customer information and supports the call center team in customer interactions.

The new reality in the Telco industry is that the basic currency of the smart network is DATA. The move to all-IP networks and the technology that has become available means that operators can collect more data than ever before from all points between their core networks and their end users and exploit it in ways not previously imaginable. Excellence in IT architecture is fundamental to efficiency and effectiveness, touching every aspect of a telco's business performance. Fortunately most senior

Telco execs have already realised that the integration of information systems, collecting, consolidating and making available all data efficiently is an essential requirement to ensure the viability and competitiveness, avoid errors and waste, improve efficiency and increase the success factors internal. As such any strategic plan to transform the BSS/OSS using EA must :

• Align the needs of information systems with business strategy,
• Monitor the rapid evolution of Information Systems,
• Rationalize and monetize investments in Information Systems
• Prioritize solutions to develop in the future according to the business strategy defined by the company
• Controlling the proliferation of systems / applications isolated and walk to the integration and overall management of Information Systems

The IT industry has embraced the concept of a Service-Oriented Architecture (SOA) as a standardized, more efficient way to build enterprise IT infrastructures. I believe that SOA, together with a revised enterprise business process, is the right way to build BSS and OSS applications, because it supports more agile internal operations, enables interoperability among new applications, and can be used to leverage existing BSS and OSS assets by adapting them to the SOA model. However to yield genuine value, an architecture transformation also requires a substantial shift in mindset.

Mckinsey are right about the fact that transforming a large telco's enterprise architecture management function to deliver maximum value is a Herculean task and multi-year effort requiring full buy-in from the business side. There is no uniform panacea for success. But the impact on costs and business performance can be huge once the enterprise architecture moves toward a uniform blueprint with consistent management across domains. Tariff changes take days rather than months. Customers can be tracked across their lifecycle and targeted with optimally customized offers, while network utilization soars.

---♠---

2.3: Telco Transformation in a nutshell

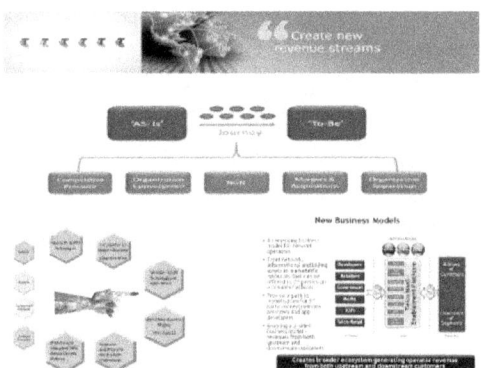

Most Telcos have recognised that Mobile Broadband is the biggest growth segment in the telecommunications market in Africa since the shores of the continent are awash with submarine cable bandwidth. The next generation of customers expects their services highly personalized, with names and images of their own choosing, integrated with their community and able to support self–expression and viral models

Telcos confront rapidly evolving new technologies even as they grapple with network privatization, liberalization, and significant changes in the regulatory framework. While most operators have mastered their own profitability economics and subscriber value, many lack insight on the economics of adjacent or competing business models. Consequently they fail to understand how to monetise the Data Tsunami.

The current "smokestack" view of network and service management is not well-suited to the demands of today's converged networks and services. Increasingly revenues are being lost to poor systems integration and operational problems, such as database inefficiencies. These problems are contributing to increased operational costs and delays in service introduction.

Today, we see many leading operators around the world making public statements about their commitment to transform but addressing different drivers: transforming the legacy network to support , growth assets, transforming cost structure (CapEx, but mainly Opex streamlining business processes, customer experience product portfolio, focusing on investments for growth.

Some carriers are replacing large parts of their infrastructure to remain the low–cost provider. The carriers are also changing how they go about network renewal, shifting from a product to a program focus, and creating a quite different kind of platform. (BT has been a very public example with 21C). Starting from the center and working outwards towards customers, the key items marked for renewal are:

• Optical Transport
• MPLS Core
• Softswitches and Media Gateways
• Multi–service Edge
• 4G Wireless Access
• Wired Access va Fibreoptic
• OSS / BSS

Transformation means transforming from a bureaucratic to a customer-focused and market driven corporation capable of competing in a liberalized market .To create workable blueprints for guiding this transformation, we must take into account the web of inter-relationships between markets, products and services, core processes, technology and systems, organization, staffing, metrics, and services models. Telcos need their business strategies revised, increasing the emphasis placed on:

• Improving network performance and customer satisfaction
• Reducing OPEX associated with network provisioning and assurance (service creation and service delivery costs)
• Shortening time-to-market cycles

Network transformation encompasses strategic technology, architecture, and implementation plans to gracefully transition access, aggregation, routing, and transport networks to an all IP infrastructure;

The transformation program must set a clear vision for the entire organization as it improves operational excellence and moves towards the defined target architecture. Create a high-level transformation checklist that covers a number of tasks starting with problem acknowledgment, followed by executive buy-in and budget approval. The Transformation agenda is based on the following pillars :

1. Analysing the current and future technology investments from business and technical viewpoints. In addition review the key customer drivers and applications that generate fast ROI based on understanding the needs of target markets. Conduct a network inventory to identify and retire and redundant network elements

2. Investigate how to incorporate Web 2 / Telco 2 paradigms into the creation of your product portfolio. Web2.0 is the new generation of web services, characterized as more open, flexible and participatory in terms of creating content, applications and collaborative alliances.

3. Strategise capabilities to overcome OTT net players to make money from higher value added services by implementing " smart pipe "design. We must assess every aspect of the network from its underlying hardware and systems to its configuration, capacity, traffic flow, and survivability.

4. Perform a thorough evaluation of the operations, identify the gaps between the current methods and the future vision, define new job functions and processes, prepare a road map for transformation, and facilitate the migration process.

5. Streamline the architecture with the judicious use of web services and services-oriented architecture (SOA). In addition to streamlining the network management systems environment, platforms and tools that enable service and customer management need to be introduced.

6. People and human resource skills must be upgraded to meet the needs of the new organization structure and new IP based technologies .Nimble, efficient operations rely on modern business processes, management practices, and human resources

7. Have a clear view on the risks and formulating mitigation strategies. Risk assessment must extend beyond the usual financial and regulatory risks to consider the wider environment in which the organization operates and the full extent of its operations, now and into the future. A failure to shift the business model from minutes to bytes or misunderstanding the changing customer mindset and insufficient insight into latent data assets are such risks.

To ensure a controlled business transformation, the transformation elements need to be planned and executed holistically using Key Business Objectives as a continuous guide. The Balanced Scorecard has three major categories of metrics which can be applied to Telcos:

• Revenue and margin: providing a view of fiscal performance
• Customer experience: providing a view of the measures that impact the end-customer's reaction to the service offering, and thus drive loyalty
• Operational efficiency: providing a view of cost and expense drivers

In Africa the smaller Telcos are constantly seeking investment capital to expand their networks for greater capacity , throughput and coverage. Some of the issues to be resolved before the investors put up cash relate to supply chain economics , the effect of nex gen telco services on financial performance , 4G monetization strategies , Corporate Governance and the calibre of the management team.

Spectrum assets alone does not cut it anymore. They must transform and streamline their operations in order to attract investment capital

--♠--

2.4: Converged Billing Systems : 4 G World's killer technology

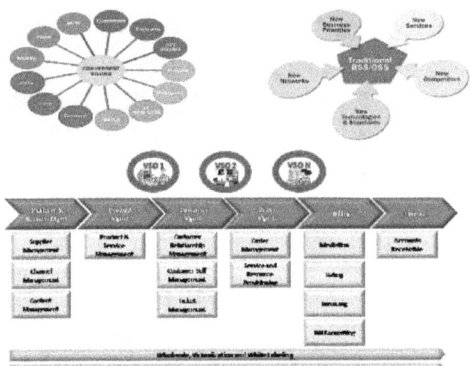 Telcos today are facing margin pressures through more intense competition, ARPU erosion, customer churn and cost issues. While designing new business models, Telcos can leverage their network capabilities, such as mobility, messaging, location, presence, profile and call control, and combine these with internet-style services such as social networking, search, advertising, direct marketing and mapping, thereby enabling richer, more compelling and more personalised services than the Internet players can offer.

Furthermore, by exposing these capabilities in a secure, controlled and automated manner, Telcos can generate revenues

from selling service enablers, as well as their own services, allowing them to fully exploit their network assets.

In light of the above Telco Execs need to understand that :

• The design and bundling of applications, content and devices to generate revenue from broadband networks is based upon a deeper understanding of the customer's data consumption habits

• The business and technical logic underlying services delivery platforms because telecoms networks have evolved from voice-centric "legacy" technologies such as SS7 and IN towards data and multimedia-centric technologies based on IP, such SIP , Daimeter and IMS

• The critical role of converged billing and CRM engines and how to convert BSS/OSS into revenue generating assets and the need to introduce attractive, profitable new services to subscribers with minimum time-to-revenue while controlling costs

Telcos can greatly benefit from implementing convergent customer care and billing systems because investing in a new stovepipe billing system for each type of service is an expensive and obviously sub-optimal proposition. The systems would help them bring new services to the market quickly, enabling them to improve customer loyalty and reduce customer churn, especially in an environment, where customers jump from provider to provider to get the best deals.

Given the challenges operators are facing with their legacy charging and billing infrastructure, they must evolve it if they want to create profitable next generation services, effectively monetize the rising demand for data, and generate revenues from third party application and content providers that are emerging as competitors. Both the customer management system and the billing system contain vital pieces of customer data. It is only when these two applications are tightly linked, and provide an assembled vision of the customer, that a superior customer experience can be achieved.

An ideal billing platform has a multi-tier architecture. Each tier communicates with other tier through published API's on the TCP/IP protocol. The Network layer constitutes a range of network devices including Access Servers, UMTS, GGSN and many more such devices.

• The Mediation layer constitutes the Provisioning server, RADIUS and Mediation Engine. Components at this layer interact with elements in the Network layer via varied protocols such as SOAP over HTTP/XML, AAA, SNMP, FTP/FTAM. Mode of exchange can be in real-time or in batch mode and can be through push or pull techniques. A new device added in the Network Layer enables services by adding a new plug-in in the Mediation and Provisioning layer. All components at this layer can run on multiple instances and also on different machines. Thus it facilitates billing to be distributed as well as scaleable.

• The Billing Platform is the central hub of the overall architecture which interacts with every layer through published APIs on TCP/IP layer. The Billing platform constitutes Billing, Rating and the Database store. The Billing platform interacts with mediation through published APIs. This gives tremendous benefit of being open and expandable.

• Rating collects formatted xDRs from Mediation layer, filters them, correlates them, applies rates, usage discounts, promotions, submits rated usage records to billing engine. Rating can run on multiple instances and also on multiple machines for scalability requirements. Rating can be done in batch or real-time.

• Billing takes input as rated records, applies taxes, discounts, calculates charges and finally generates invoice records for every customer. Billing can take rated records from 3rd party rating engine, 3rd party billing system and can also exchange invoice records to 3rd party systems like FAS, Billing Printing. All these are possible because of well published APIs of Billing. The Billing Engine can also run as multiple instances and in multiple servers.

• Customer Care includes, CRM, Customer Order Management and interface to external messaging systems. This layer is also independent of other layers of the system. Thus CRM as a whole layer can be in public LAN which is direct interface to customer and call centre. Customer care interacts with billing, customer management, 3rd party messaging system through well published API's.

In both the wired and the wireless arena, convergence offers an opportunity for the service providers to differentiate themselves from their competitors through creative bundling of services and billing strategies. Which in turn would increase customer loyalty. Thus, as service providers add newer services to their offering, it becomes imperative for them to develop an integrated billing mechanism.

As complex as it may sound , billing in the context of convergence is one more crucial piece of the strategic puzzle telecom service providers need to grapple with as they evolve into strong marketing organizations in a hypercompetitive world.

--♠--

2.5: The Digital Telco : Telefonica is the mighty Sensei !!

Digital commerce is evolving fast, moving out of the home and the office and onto the street and into the store. The advent of mass-market smartphones with touchscreens, full Internet browsers and an array of feature-rich apps, is turning out to be a game changer that profoundly impacts the way in which people and businesses buy and sell. As they move around, many consumers are now using smartphones to access social, local and mobile (SoLoMo) digital services and make smarter purchase decisions. For most telcos, the best approach is to start with digital commerce, where they have the strongest strategic position, and then use the resulting data, customer relationships and trusted brand to expand into personal cloud services, which will require high levels of investment. This is essentially NTT DOCOMO's current strategy.

Communications service providers have a number of attributes that give them a potential marketplace advantage in the Digital world of commerce and comms : an extensive customer base, distribution muscle and knowledge of customer preferences through CRM and billing systems. The opportunity is to become an integrated digital services provider across platforms and mobile devices—convincing customers that a communications service provider can effectively serve as the hub to meet their communication and entertainment needs.

According to an Accenture survey, the areas that show particular promise include cloud services and location-based offers. In May, 2014 Norway's Telenor announced a new Group strategy and Digital Unit in order to boost its responsiveness to customer needs. But there is one Telco that has taken the notion of Digital imperative to stratospheric levels : TELEFONICA.. my favourite Telco Titan !!

Telefónica has been rapidly transforming into a Digital Telco par excellence. They aim to provide the digital products and services which will help to improve the lives of our customers by leveraging the power of technology. This ranges from developing new technologies for consumers to communicate with friends and family through to helping businesses and governments address new opportunities, improve operations and increase efficiencies.

As of February 2014, Telefónica implemented a new organizational structure, one that is completely focused on clients and incorporates this digital offering as the main focus for commercial policies. The structure gives greater visibility to local operations, bringing them closer to the corporate decision-making centre, simplifying the global structure and strengthening the transverse areas to improve flexibility and agility in decision makings.

So that's a global operational revamp and a challenging new networks strategy — a lot of change within a large global organization (about 130,000 staff and 323 million revenue-generating connections). Within this framework, Telefónica has created the role of the Chief Commercial Digital Officer (CCDO), who is responsible for fostering revenue growth. On the cost side, the Company has strengthened the role of the Chief Global Resources Officer. Both Officers will report directly to the Chief Operating Officer (COO), as will the local business CEOs for Spain, Brazil, Germany and the United Kingdom, in addition to the Hispanic-America Unit.

The reorganization comes at a critical juncture for Telefónica, as the operator is also embarking on a radical evolution of its network infrastructure with the introduction of NFV and, ultimately, SDN capabilities. "capture gross savings of up to €1.5 billion "in the next year", as well as having responsibility for "the synergies plan in Germany.Value-added services are the competitive edge, but the strategy – and how deep into the water a Telco goes in placing itself at the centre of the Telco 2.0 environment is very important as well. While other companies take a "defensive stance" focusing just on value-added services, Telefónica has fully embraced the multi-sided business model, acting as a middleman between end customers and other third party companies, according to a study by Telco 2.0 Research.

What are the technologies that will help communications service providers such as Telefonica , SingTel , Telenor etc meet current needs and the challenges of the future? According to Gartner, CSPs are actively considering deployment of a new breed of IT-centric services that involve the use of data, analytics and digital content. CIOs will need to transform business-only, support IT to effectively deliver customer-facing digital services.To transform CSPs into diversified service providers, Gartner said that CIOs must harvest business value from existing networks, IT and information assets, and hunt for new opportunities using digital services.To drive new revenue, CIO offices must acquire new skill sets to align IT to business strategies.

The priorities of telecom CIOs are related to business applications, BI ,analytics, customer service support systems, centralizing the billing systems and the convergence of Internet protocol. While CIOs work to deliver applications and innovation, they are also asked to be as cost effective as possible. Looking ahead, IT leaders from communications service providers need to be aware of the shifts brought by new technologies – and they will be requested to provide insights on how carriers should provide strategic plans and new services to be launched based on them.

This may involve changes to the infrastructure layer to fit new demands.

You recall ofcourse the seminal IBM Whitepaper Telco 2015 that painted4 future scenarios for Telco players. One of the scenarios is called " Generative Bazaar ". In this scenario pervasive, affordable, open connectivity is enabled for a person, device or object, unleashing a wave of generative innovation. A co-operative of horizontally integrated network infrastructure providers (Net Co-op) emerges, based on catering to the needs of a multitude of asset-light service providers that package connectivity with completely new services and revenue models.

IBM modeling of its four scenarios suggests Generative Bazaar as the most attractive outcome in terms of revenue, profitability and cash flow projections. This is precisely what Telefonica's Digital dream is all about.Telefonica takes an entrepreneurial approach to identify and accelerate the latest technology trends, incubating new disruptive digital businesses (Wayra and Amerigo). They aim to deliver end-to-end digital products and services in B2B, B2C and B2B2C in a number of domains: future communications, machine-to-machine, financial services, media services, cloud computing and information security. They also develop the right partnerships so they can be the single source of the best digital experiences for its customers.

Telefonica believe its strategy is supported by "four pillars in the short term": growing revenue by extending the commercial offering to new services in the digital world; modernising networks and systems, through accelerating the deployment of the most modern technologies; increasing efficiency through simplification and cost-cutting as well as ongoing financial discipline, prioritising investment in growth projects that generate added value; and strengthening its leadership in the digital ecosystem, "by driving a new public positioning enabling the hypersector to re-establish balance in the value chain".

In essence Telefonica has reorganised itself to create , catalyse and capitalize on opportunities along the following vectors : • Innovation & New Business – focus on discovering and incubating the next generation of digital services. They take an innovative 'venture capital'-like approach to innovation, as well as identifying the right start-ups to work with, invest in or potentially acquire. Their crowd data offering, Smart Steps, sits within this unit.

• Digital Services – delivering high quality, integrated product offerings across a number of more established digital service areas, including Machine-to-Machine, Cloud Computing, eHealth, Financial Services, Advertising and Information Security. It also brings together all of Telefónica's capabilities in media services ranging from IPTV and satellite TV to over-the-top video, ensuring that Telefonica is a leading video company. • Communications – enabling Telefónica to continue innovating in its core business of communications, leveraging its world-class expertise in IP and video communications.

Customer ownership and distribution power give CSP's a strong foundation on which to build to meet consumers' ongoing communication and entertainment needs. Providers have an opportunity to improve their return on investment by monetizing better connectivity. They can also extend their partnerships across the digital ecosystem to provide a seamless customer experience.

This will require deep insight into subscriber behaviors, new forms of collaboration within the industry, new capabilities within the organization and an ability to constantly innovate to keep pace with today's demanding consumers. They must establish value proposition for third-party providers, including interfaces to network capabilities, service enablers based on open standards, access to ecosystem of partners, a commercial model and infrastructure support for common business process services, e.g.: self-service, e-commerce, billing Leaner operating models and new leadership roles designed to spearhead digital growth are a critical success factor for Telcos, particularly those with large footprints incorporating markets at different stages of maturity.

Success in the IBM Generative Bazaar scenario is dependent on the provider's ability to achieve structural industry separation; a pervasive open network access infrastructure; support for third-party application / services innovation; a dynamic business design; and the ability to leverage advanced customer and network analytics.

By 2016, the fastest-growing markets will add nearly 1 billion new mobile connections and account for 56 percent of all mobile connections worldwide. While these hypergrowth markets will be the global driver, they also pose challenges that go beyond cross-border execution and cultural conflicts. CSPs will need to transform themselves into leaner, more industrialized digital companies. And Telefonica is the undisputed global leader in the digital endeavour.

--♠--

2.6: Africom 2013 : Solving the " Cost efficient " networks puzzle

Last week the who's who of the Telco zoo gathered in herds at the 16 th annual Africom Event in Cape Town. Yours truly moderated the keynote session : COST EFFICIENT NETWORKS which highlighted the requirement for increased investment in the networks, to cope with the data explosion and identified strategies for increasing the quality of the network to the highest standard without excessive spend. Delegates in the stream focussed in how to reduce capex and opex in a data hungry environment since pressure on prices and margins is a situation faced by almost all operators. The telecom scenario shows a world going "flat" with abundant voice and data volumes while the cost of promotional discounts to attract new customers increases : the disconnect between traffic and revenues !!

In MEA , fixed and mobile Telcos have not fully realized the cost-reduction potential provided by lean tools and techniques, which not only can generate savings of from 10 to 15 percent on the addressable cost base, but also simultaneously improve overall operational quality levels.This process should start with a diagnostic phase that covers network planning and implementation, operations, and management infrastructure. There are also a broad range of new OPEX saving possibilities which can be leveraged through a new generation of technologies, e. g. in the area of software defined radio networks (SDR) and self organized networks (SON).

Several speakers from the satellite industry made impassioned presentations without really explaining how satellite fits into the creation of a cost efficient network in Africa. Unfortunately vendors get myopic in their desire to punt technology solutions instead of anchoring their presentations on the theme of the stream and deliver presentations that have educational merit. Atleast the panel discussion elicited useful points such as Green Telco initiatives as well as concrete ways to reduce capex/opex. Going green is in many ways similar to going lean, since both strive to reduce energy usage and eliminate waste.

Several Telcos have gone public with energy efficiency,power reduction, and carbon footprint reduction objectives. Verizon has established an objective for its vendors to achieve 20 percent greater efficiency by January 2009, as compared to today's equipment. France Telecom is planning to reduce the greenhouse emissions per customer by 20 percent between 2006 and 2020 and British Telecom claims to have reduced its carbon footprint by 60 percent since 1996, and has an objective to reach 80 percent by 2016. Fixed and mobile operators can foster green networks by improving network energy and cooling infrastructure, and by installing energy-saving network equipment.In terms of energy operating costs, operators must initiate active programs to identify sites with higher-than-normal power consumption and adopt specific measures to reduce it. This can include adjusting air conditioning settings, making productivity upgrades to batteries and A/C systems, and adopting low-energy designs. Companies must also investigate the transfer of expensive third-party energy contracts to players that offer better terms and conditions.

The Gurus at McKinsey believe that even that fixed-line infrastructure players will outsource network infrastructure and operation to contractors in order to optimize operating and capital expenditures (opex and capex). Making this outsourcing a success requires companies to explicitly split roles and responsibilities with the chosen contractors, establish clear reporting and interface models, and prepare, negotiate, and execute specific contracts and service level agreements.

Telecoms players can employ proprietary analyses and techniques to improve the amount of value their products deliver to customers, while at the same time, creating cost-efficient designs and calculating target costs.While personnel wages and benefits represent a major network operating cost, other high potential areas for cost cutting include site rental and energy costs. As a consequence, some operators are aggressively pursuing the renegotiation of rental contracts with an eye toward moving or eliminating those sites with the most expensive rental contracts. Considering network optimization, some operators are exploring base transceiver station (BTS) "hotels." These BTS hotels group the electronics from a number of base stations for antennae up to 20 km away.

According to the luminaries at Arthur D Little ,full cost transparency must be established in order to identify, prioritize and optimize saving measures. For this reason the first phase of the cost reduction project needs to focus on establishing a stringent OPEX/CAPEX analysis. Operational saving measures are considered as activities which render saving benefits of typically in the range of 10% p.a. These measures include obvious examples like the change of maintenance service level and backhauling optimization or reduction of product portfolio. Less obvious cost saving measures include the introduction of QoS concepts for optimized bandwidth management, reduction of room temperature in local exchanges or the ceasing of 3rd party hardware maintenance for stable legacy infrastructure where better maintenance know-how is often already available in-house.

The network operations centers of many telcos face a variety of challenges, including having to deal with technology silos, unclear ownership of network issues, lack of institutional memory that forces teams to "reinvent the wheel" time and again, and others. Given the breadth of opportunities available, operators can often capture reductions of 15 to 35 percent in NOC-related costs. Potential actions include developing a clean-sheet NOC redesign, integrating NOC services on an end-to-end basis, and instilling a problem-solving, high performance mindset within the center. By introducing optimized governance models, best-practice vendor relationship management techniques, and better negotiation and deal strategies, operators that revisit mobile outsourcing typically identify the potential for an additional 5 to 10 percent in cost reduction, representing 2 to 3 percent of total costs.

Mobile operators can make use of the rich variety of customer data they have on hand to improve their network quality and target investments on a site-by-site basis. Taking this type of highly granular review of network performance metrics, site utilization, and commercial performance will enable leaders to pinpoint spending requirements.By using techniques such as network caching and CDN, operators can reduce one-on-one network downloading and hence, network load and form partnerships with broadcasters to share investments and build a large-scale, secure, single network infrastructure.

Revisiting the organization's zero-based budgeting decisions using the latest insights and business priorities can reveal new opportunities to reduce investments andcosts in areas where an operator's market share is below critical thresholds. Competitive pressure is eating into revenues and causing a spike in subscriber acquisition and retention costs (SARC). The e Channel is an opportunity in cost reduction since it enables signing up higher ARPU subscribers at lower cost of sales and cost .Key success factors for an effective e channel strategy include : create a compelling channel experience (exclusive offers) ; build a solid IT infrastructure (SOA) and revamp the organizational structure (self-contained empowered eChannel teams).

Customer self-care has been shown to reduce costs in customer contact centres by as much as 20%. This sector has evolved to become a full set of self-service capabilities that includes customers researching and buying through self-directed channels. Buying via these means has been shown to increase the revenue per user by as much as 18%, when the Telco provides an effective self-service interface for the customer.

A basic cost reduction mechanism and culture across all staff must be in place (e. g. personal target setting, cost transparency, etc.). The challenge is to embed an organisational discipline that will constantly challenge the existing cost basis. The benefits of creating a performance-driven culture within Telcos come from its capacity to amplify subsequent improvement initiatives – in effect, supercharging them. However, as with most transformational approaches, "getting there" requires strong, visible commitment from company leaders, solid organizational planning and training, and communication clarity.

One thing for sure : solving the cost/efficiency puzzle requires a wholistic mult facet approach that will target the right levers to optimise cost even as capex is injected into building high speed IP based broadband networks.

--♠--

2.7 : Telcos and Big Data : Ride it or drown in it : What's your gameplan ??

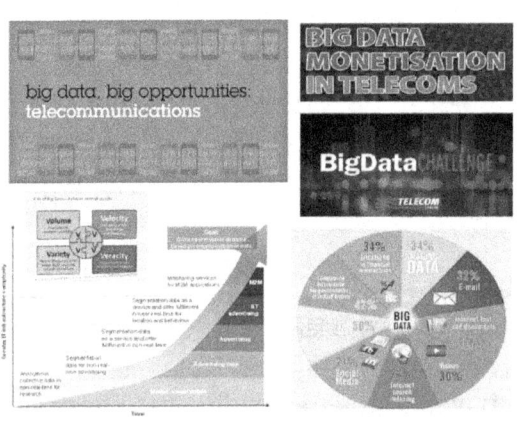

Big data has been a headline theme in the technology and mobile space for some time.Telcos all over the globe are seeing an unprecedented rise in volume, variety and velocity of information ("big data") due to next generation mobile network rollouts, increased use of smart phones and rise of social media. Telco operators have historically focused on managing the network with little visibility to the impact it has on the customer's experience. Which means the operator was forced to work with snapshots of network data in fragmented views or at a summary level in order to plan network capacity or provide information to customer care and marketing about customer transactions until now !!

Big Data technologies, and in particular their analytics abilities, offer a multitude of benefits to telecom companies including improved subscriber experience, building and maintaining smarter networks, reducing churn, and generation of new revenue streams. Mind commerce, expects the Big Data driven telecom analytics market to grow at a CAGR of nearly 50% between 2014 and 2019. By the end of 2019, the market will eventually account for $5.4 Billion in annual revenue.

Mobile commerce is one particular area where operators and service providers can potentially deliver tangible benefits from the application of big data analytics. The growth of m-commerce is creating large amounts of information on consumer behaviour and choices, which can be used to offer more personalised services and offers. SK Planet (Division of SK Telecom) have stated that "our Cash Bag m-commerce portal should generate $9.3 billion in revenues this year, and by using big data analysis we can provide customers with a much improved experience, and not based simply on offering the lowest price."

Big data analytics solutions enable service providers to analyze real-time location data over time for opt-in subscribers to understand subscribe lifestyle. Combining lifestyle and mobile profiles with subscriber usage and digital behavior allows service provider to create targeted offers for opt-in subscribers. With a majority of subscribers using smart phones to access data services as well as voice, mobile network operators are seeing explosive growth in traffic levels across their networks. In addition, the mobile network operator environment is fiercely competitive, with the ability to attract, retain and grow valuable subscribers being key. Increasingly, the provision of high quality customer care is an important component in the marketing mix and in retaining subscribers.

The growth of connected devices, particularly in areas such as the home or in the car, presents new opportunities but also challenges for operators and other ecosystem players. Users may be willing to share data with service providers but on the basis that the data is used securely. This year the GSM industry introduced a standardised mobile identity solution that aims to become the de facto single sign-on tool that consumers could rely on to authenticate themselves in both online and offline environments. This initiative is set to stimulate adoption of mobile services that rely on absolute confidentiality, such as healthcare, government and banking.

The announcement of 'Mobile Connect' has came alongside a number of other partnerships are aimed at accelerating the uptake of digital commerce services via mobile solutions. Mobile Connect is backed by 12 leading operators including Axiata Group Berhad, China Mobile, China Telecom, Etisalat, KDDI, Ooredoo, Orange, Tata Teleservices, Telefónica, Telenor, Telstra and VimpelCom, as well as key industry players such as Dailymotion, Deezer, Gemalto, Giesecke & Devrient, Morpho, Oberthur and VALID.

The Open Mobile Alliance is also developing standards and specifications for how devices and apps should share data with other connected devices, supporting the development of interoperable end-to-end mobile services. The move towards increasing standardisation is a key element in reassuring consumers and addressing concerns of privacy and security around personal data. The proliferation of smart phones presents new opportunities and challenges: consumers want the best deals for all purchases based on their real-time location while requiring the services provider to honor their privacy preferences and provide only relevant offers when requested/opted-in.

Given the highly secure capabilities of the SIM, mobile phones could become the perfect tool for future Digital Identity, not only for digital use cases but also for authentication in offline environments (national ID card, airport check-ins, etc). This makes

Mobile Connect initiative ideal in a range of environments (both online and offline), yet this web-based authentication service does not necessarily need to be linked to a SIM to function.

Mobile Connect is a web-based authentication service runs on the OpenID Connect protocol, ensuring interoperability across mobile operators and service providers. The identification solution being developed will use the subscriber's mobile phone number or mobile user name and information contained in the secure SIM card, meaning that consumers will no longer need to create and manage multiple user names and passwords.

At a strategic level, the Mobile Connect service establishes the SIM card and the mobile medium as a frontline identity management service provider, allowing mobile operators to participate in the critically important e-commerce market. In the longer term, this type of standardised mobile identity solution will help operators to derive substantial revenues from their presence in fast- growing e-commerce markets, notably by extending the reach and presence of operators' brands, raising levels of awareness and ultimately, improving loyalty.

Big Data opens a vast array of applications and opportunities in multiple vertical sectors including, but not limited to, retail and hospitality, media, utilities, financial services, healthcare and pharmaceutical, telecommunications, government, homeland security, and the emerging industrial Internet vertical. In fact, according to Heavy Reading's Big Data & Advanced Analytics in Telecom report, the industry will move from the $1.95 billion it spent on Big Data and analytics in 2013 to $9.83 billion in 2020. Thats really big business even if we are to take these projections with a pinch of salt. Ofcourse you will have to hire the right set of skills to make sense of and monetise the data deluge.

And while solutions are already present that will help you ride that wave of data, the question is: how can you profit from it? Here are some insights into how telecom companies can use the power of Big Data to their advantage.

1. Monetize it : Brands like Telefónica O2 are figuring out how to provide valuable analytical insights to customers to help them become more effective. Using mobile network data, telecoms provide valuable details into shopping habits.

2. Sell Subscriber Insight Data : Other companies like AT&T are looking at selling aggregated customer data to marketing and advertising firms.

3. Advertise Smarter on Mobile : Still other brands are using Big Data to strengthen their mobile ad campaigns. SingTel recently acquired mobile advertising platform Amobee, in an effort to help clients better reach their target audience and deliver more relevant offers.

4. Better Analyze Set Top Box Data : Using the data that today's sophisticated set top boxes offer may provide new revenue streams from targeted ad sales and more customized and personalized content services.

The Gurus at Strategy & believe that many types of data are potentially available to operators — and certain sets of data might be combined to open up new business opportunities in areas such as campaign marketing and fraud prevention. Operators could generate more accurate and personalized offer recommendations for existing individual subscribers by combining internal structured data, such as how and where each subscriber uses his or her phone, with external unstructured or semi-structured data from social media platforms (for example, Facebook and Twitter).

By correlating internal location, usage, and account data with external sources such as credit reports, operators could significantly increase the detection of fraudulent activity such as looping or call forwarding on hacked PBXs (private branch exchanges), or fraud involving the swapping of SIM cards, and improve the overall accuracy and efficiency of their efforts to recognize patterns of fraudulent behavior.

Imagine having the best of both worlds ? Having the tools to analyze the growing amount of data and content your business is generating, and also finding ways to make it profitable. If you are astute then this deluge of data isn't a threat; it's a serious opportunity to take your telecom business in a new, exciting, and yes, profitable direction !!

--♠--

Chapter 3 : The road to 4G monetization

3.1 : IPX and LTE Roaming : A Telco Wholesale Opportunity

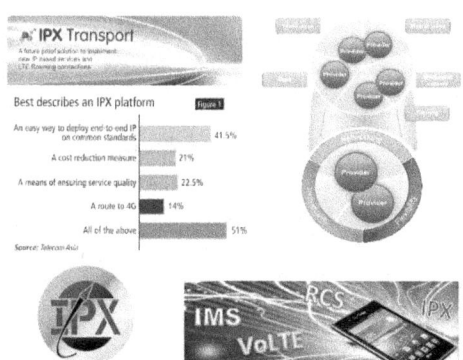

Recent months have seen an increasing number of wholesale telecoms operators and other companies enter the market with inter-carrier IP exchange (IPX) services. Several developments will drive the long-term demand for IPX services, including the launch of 4G networks based on

LTE, the growing demand among VoIP service providers for high-quality transit and termination, and the ability of IPX to support new services, including videoconferencing and HD voice.

By providing hub-based interconnection to telecoms operators, content providers and other companies, IPX service providers aim to offer a private international network for exchanging both IP and legacy traffic that is separate from the public Internet. As much as 25% of the wholesale value of traffic flowing over IPX networks is expected to be generated by new and value added services, with LTE interoperability, roaming and voice services set to be strong drivers for growth, according to telecoms analyst house, Innovation Observatory.

IP exchange (IPX) is a telecommunications interconnection model for the exchange of IP based traffic between customers of separate mobile and fixed operators as well as other types of service provider (such as ISP), via IP based Network-to-Network Interface.The intent of IPX is to provide interoperability of IP-based services between all service provider types within a commercial framework that enables all parties in the value chain to receive a commercial return. The commercial relationships are underpinned with service level agreements which guarantee performance, quality and security.

French operator group Orange's wholesale division has expanded its IPX (IP exchange) with a Diameter signalling offering allowing customers to introduce LTE roaming across Europe, the Americas and Asia.The operator said that it has an ambitious program in place to extend it's LTE Signalling capability through direct connectivity and peering agreements. According to the operator the Diameter-based LTE signalling service enables operators to provide end users with improved QoE on 4G networks while roaming outside their home country.

The Telenor Global IPX enables operators and partners to be connected through one IP connection. The IPX interconnect solution allows for an optimized, flexible, secure connection and guaranteed Quality of Service (QoS) for voice and mobile services. This is an important shift to improving Global Roaming Quality for the end customer. The Telenor Global IPX Service is provided over the IPX Compliant MPLS/IP Network and includes the basic framework designed by the GSMA which incorporates four main principles:

• Premium quality
• Secure environment
• Flexible for all services
• Cascading payments

To establish LTE data roaming worldwide, two architectural issues need to be addressed. First, it's not that easy for an LTE operator by itself to complete interconnectivity with its global roaming destinations. Instead of current GRX and bilateral TDM links, roaming between LTE/EPC networks requires an efficient international exchange mechanism to integrate all the IP-based services originated in EPC/IMS and to interconnect them among numerous LTE operators worldwide. Besides, the existing GRX defines only best-effort IP transport, a QoS guarantee scheme between visited and home networks is required to let LTE subscribers enjoy international services with the same quality as domestic services.

Telstra Global is developing its capability to become a leading IPX provider in Asia-Pacific on a base of service flexibility for LTE roaming and service-aware traffic management. Telstra Global's IPX strategy for LTE roaming borrows lessons from the cloud by offering its customers the ability to burst traffic in peak times. Telstra Global is also supporting the dynamic allocation of classes of service (CoS) to address the increased requirements of the content- and applications-based traffic expected to flow over next-generation networks. Telstra Global sees basic LTE data roaming as just the beginning of its IPX ambitions

Second, instead of the legacy SS7 MAP in GPRS roaming, roaming signaling in LTE/EPC uses Diameter on IP to exchange subscribers' authentication and location update between visited and home networks. Diameter is the Authentication, Authorization and Accounting (AAA) protocol succeeding to RADIUS. Unfortunately, interoperability of Diameter technology hasn't yet reached full maturity. Actually, different vendors' EPC nodes sometimes can't successfully communicate to each other due to different implementations of Diameter signaling despite 3GPP standards.

Spectrum fragmentation is the big elephant in LTE roaming room .3GPP identifies more than 20 LTE Frequency Division Duplex (FDD) frequency bands and more than 10 LTE Time Division Duplex (TDD) frequency bands. According to Analysts, LTE operators have committed to launch in at least 13 bands, and with as many as 10 bands being proposed in a single region. It increases the cost for LTE operators to provide global roaming coverage with multiple bands, as well as for manufacturers to develop roaming-compatible handsets.

To solve this problem, the industry has to seek a set of common bands for international roaming or has to wait for a handset that supports a sufficient amount of frequency bands. Whereas the currently popular bands (800MHz, 1.8GHz and 2.6GHz) are likely to be applied for international LTE roaming for the time being, a global consensus on common bands for international roaming is necessary for sustainable growth of LTE and its roaming.Despite the spectrum problem, a lot of MNOs, IPX providers, and vendors are preparing and testing LTE roaming now.

Therefore, international LTE roaming will certainly expand particularly in the popular frequency bands. Thus, international

LTE roaming on IPX embodies interconnectivity of IP-based telecommunication services among LTE/EPC ecosystems and reflects the beginning of new experience on global coverage of high speed mobile computing.

Recently SAP AG announced an IPX peering agreement with Etisalat UAE, the largest telecom operator in the Middle East and Africa, to deliver LTE roaming traffic to all mobile operators. This strategic LTE roaming and diameter peering agreement will help Etisalat operator companies to interconnect with SAP Mobile Services' strong IPX customer community and launch LTE roaming quickly.

---♠---

3.2 : Monetising LTE : A review of savvy pricing strategies !!

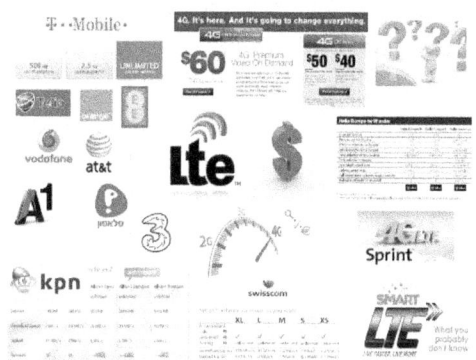

Whilst LTE is spreading across the globe,(222 commercial LTE operators in 83 countries GSA Evolution to LTE report: October 17, 2013) the question remains how to best leverage the speed and capacity that LTE brings to create value for customers and generate more revenue.).The introduction of LTE should be used to enhance the value offered to consumers. Operators should introduce new pricing schemes (e.g., leverage quality of service differentiation), enter new markets (e.g., fixed broadband, services such as video streaming and calling), and gain market share (e.g.,in mobile broadband and in sub-segments such as the high-end customer segment).

Based on improved technical features and economics, operators can selectively utilize customers' greater willingness to pay to counter the price reduction trend. New mobile Internet customers need to be educated on the superior service quality on offer and drawn to the brand with introductory offers to counteract low willingness to pay.

With LTE, tiered pricing is evolving from volume-based tiers to speed-based tiers with data packages based on varying speed entitlements, but often a combination of both.Bear in mind that implementing speed-base packages requires sophisticated policy control architecture connected to an online charging system. Typically when a subscriber begins a data session it triggers a message to a Policy Manager which reviews the Subscriber Profile Repository for the customer's entitlements. The Policy Manager enforces the rules through the Enforcement Point (a network packet gateway or deep packet inspection node). Policy management can play a significant role in this initiative by providing operators with a way to offer their customers innovative, personalized bundles of services that are consistent across multiple channels. Via initiatives such as cross-sell/upsell, highly targeted services and creative packaging, operators can provide their customers and prospects with more value and, in doing so, drive loyalty behavior

There have been a lot of discussions around unlimited data offers, with many analysts arguing that they were unsustainable and that operators needed to create more differentiated and value-based offers. Swisscom has actually been using speed as the differentiating value whilst offering unlimited data on LTE – with its Infinity tariffs launched in July 2012. Based on the fact that the value of speed is easier for customers to grasp than data volumes, this approach gives customers the worry-free "unlimited data" that they like, whilst still differentiating to optimize revenue.

As a result, Swisscom has seen its overall ARPU grow, although it initially declined its market share is increasingly growing. KPN Holland are packaging speed as the main promoted value of LTE. It enables them to create differentiated packages where speed is used to create perceptible value – and encourage customers to upgrade to higher tiers and 3G customers to switch to LTE for a premium.

3 Austria follows the same strategy as Swisscom with its Hello and Hello Europe tariffs that include unlimited data with differentiated speeds and mobile TV channels. However, whilst offering unlimited data, both Swisscom and 3 Austria restrict the speed to 64Kbit/s once a certain amount of data is consumed; 3 Austria clearly highlights this to its customers and shows for each tariff the full speed data volume in addition to the maximum download and upload speeds. However, 3 Austria also offers a premium package with unlimited and unthrottled data speed.

Mobile US for instance offers high speed data volume tiers with its "Simple Choice" plans and then throttles to 2G data speed once the high speed data allowance is exceeded, until the next billing cycle. In effect, this approach is equivalent to offering unlimited data to customers at differing speeds. Orange France also throttles the speed when the data allowance is exceeded. Optus in Australia automatically moves customers to the next tier for the rest of the month. AT&T sends data usage alerts to its customers and then charges when they go over their allowance.

In terms of roaming, T-Mobile US offers unlimited data while roaming as part of its Simple Choice plans; however the speed is limited to 128Kbit/s , the aim being to upsell "speed packs" when higher speed is required. The plans are subject to fair usage rules and also include unlimited SMS while roaming. This approach is very attractive to customers whilst enabling the company to continue to drive roaming revenues by upselling speed. Temporary speed boost offers may also bring additional revenue from travelers by allowing them to quickly download movies or games to enjoy during their trips.

LTE has brought the opportunity for operators to upsell high-speed add-ons to supplement a base package or temporary speed boosts, enabling customers to purchase higher speeds as required. For example, A1 Telekom Austria upsells high speed options to its customers in addition to basic packages that incorporate lower speeds. A1Telekom Austria offers a data add-on (Eur 8.25) to its business customers, with night time access (22:00-08:00) to unlimited data, at full speed as per their base package. This is an interesting approach to managing speed and network resources whilst offering the always popular "unlimited" to create a new data revenue stream.

EE UK applies a speed cap of 30mbps to all new customers not on its premium post-pay package "4GEE Extra" while premium customers can enjoy speeds of up to 150 mbps. Their packages also include unlimited minutes and texts as well as free roaming calls and texts to selected countries and includes the Deezer music service. Telia, the world's first LTE operator offers its 4G post-paid "Mobil Komplett" customers a speed up to 100 Mbit/s whilst all its 4G pre-paid customers are capped to 20 Mbit/s. Telia's Mobil Komplett plan also comes with data volume tiers and shared data allowing customers to use up to 7 mobile/tablets on the same subscription. The plan includes unlimited calls, SMS and MMS.Orange France launched its low cost sub-brand Sosh which uses speed caps to differentiate from the main brand. Whilst Orange customers can enjoy LTE speeds up to 150 Mbit/s, Sosh offers a maximum of 42Mbit/s with its best package.

As the telecom market becomes increasingly saturated, operators are looking for ways to stand out without resorting to price wars. Core services such as voice or data are becoming commodities; in order to avoid being limited to charging commoditized prices, operators must be able to create and deliver services that offer additional value, but are limited by the rigidity of many legacy service creation and OSS/BSS environments.

A flexible policy management solution deployed either as a standalone implementation or as part of a larger OS S/BSS and service delivery transformation initiative, can enable value-added, differentiated services that can run on top of existing core services. While pricing can never be low enough from a consumer perspective, the ongoing quest for operators is to find a balance between competiveness and the ability to fund future investments.

---♠---

3.3 : Monetise 4g / LTE : a rational and balanced approach !!

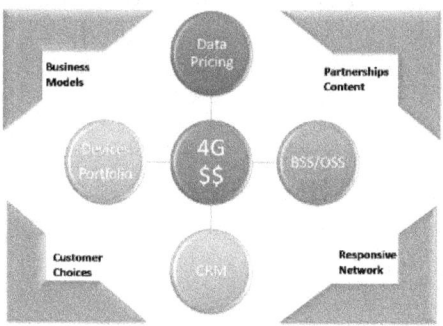

Telcos are pouring billions of dollars into building out 4G networks. LTE will stimulate demand for video and media services, thanks to its lower latency and higher capacity access, and indeed mobile TV, video calling and video downloading show the strongest growth prospects. The point is that no single nontraditional service will be able to compensate for the erosion in traditional telecom revenue. Media/entertainment (including advertising), machine-to-machine (M2M) services, cloud computing and IT services are promising areas for generating revenue.

To monetise LTE , Telcos will need to define new business models and marketing strategies to drive acquisition and retention of subscribers, as well as adoption of high-value content and applications. Optimising mobile broadband economics is a complex challenge, and there's always a temptation to try to solve complex problems with one 'silver bullet'. Unfortunately this is impossible, as there are many different combinations of solutions will work at different times for different operators.Operator device portfolios are one of the most important factors underpinning network technology migration trends.

Smartphones offering a compelling user experience drove mass-market uptake of 3G mobile broadband subscriptions and LTE will be no different. In the LTE ecosystem, tablets seen to offer a greater prospect for revenue growth than dongles. It will be important to secure an attractive selection of smartphones and devices at affordable price points.To boost the need for LTE, operators can make partnerships to enable applications that allow customer enjoy the full experience of LTE.

Telcos need to adopt a balanced approach on how to monetize 4G networks by a thorough analysis of various technical , financial and commercial strategies : the desired outcome is to minimize the cost of access coverage while maximizing subscriber capture. A balanced strategic plan is founded upon :

• Effectively reengineer the broadband business model so as to reduce costs , manage data traffic , and develop a more sophisticated approach for pricing broadband access

• Unlock new revenue streams to justify the enormous network investments over time in the context of key customer drivers and applications (cloud , M2M etc) that generate fast ROI based on understanding the needs of target markets.

• Collaborate with OTT players since LTE's all-IP architecture will create a more open environment for Over The Top (OTT) applications which threaten to further commoditize the network.

• Leverage the OSS/BSS to weaponise the CRM , Billing , Policy Control systems in order to ensure that all the data traffic is accounted for and billed to the correct entities.

What service providers need to do is to offer packages based on the service or application used that can be provisioned dynamically ,rather than on bandwidth allocation. Policy management tools play an important role here. By being able to offer management tools , the provider will be able to offer subscribers packages such as ' YouTube subscription' , or ' online gaming subscription' , or ' regular surfing and email subscription . Policy(PCRF) is the brains of a network , especially for LTE networks that must make many more real-time decisions to maintain network performance and adapt the network to the subscriber.

In a Media Research survey, respondents pointed overwhelmingly to smart devices, video, and cloud services as the devices and services most likely to drive demand for 4G. Fully 40 percent already have partnerships with content providers to assure them higher quality of service (QoS) on their networks. Enhanced enterprise LTE solutions, such as videoconferencing on-the-go and remote access to business applications, can drive data consumption. Verizon Wireless is one of many LTE operators that offers 4G mobility applications and solutions for SMEs and enterprise customers. A survey shows that 67% of US businesses using LTE believe that it has resulted in increased productivity.

In order to offer a more competitive service than the OTT players KT is leveraging CCC for ICT business. CCC is a kind of domain-specific cloud technology, based on virtualisation. By unifying the platform for radio and several application services into CCC, KT can provide cross-layer optimised services between applications and radio. For example, they can utilise user contexts such as user ID, traffic content, QoS, location, and the radio environment to provide the most suitable service to their customers.

There are some valuable options for addressing the data issue from a technical point of view, offload perhaps the most valuable amongst them. However, these are not all the weapons in an operator's arsenal. They can also look to manage the impact of traffic on their networks and their bottom lines by looking at different business model and pricing options.

Operators can use the rich data experience of LTE to sell more data and develop new revenue streams. Video streaming providers such as Netflix alter the quality of video according to available bandwidth – so a 6-minute clip on LTE would consume 80MB compared with 27MB on 3G, thus driving usage. Operators are also bundling content with LTE or top-tier plans, enabling new revenue streams – for example, EE in the UK uses its film service (EE Film) to monetise data and receives sales commissions from video-on-demand provider FilmFlex.

On the revenue side, the bulk of revenue will be from 'downstream' subscription and pre-pay customers, and while helpful, that the near-term growth of new 'upstream' or wholesale / carrier services revenues alone would not be enough to cover the costs of capacity increases.Because LTE network latency is lower than 3G, operators can develop new revenue streams by selling bandwidth for wholesale services (such as utility and M2M services). Verizon is at the forefront of this with projects in sectors such as education.

MNOs can also experiment with bundling. Data sharing across devices is being offered, with the aim of monetising devices (such as tablets) otherwise lost to Wi-Fi. Tethering strategies are evolving, as operators try to monetise tethering by allowing it at as part of premium or top-tier plans. Fixed–mobile converged offerings are available and aimed at increasing revenue and reducing churn.

There are many different possible solutions and different combinations of solutions will work at different times for different operators :

* New air interfaces and spectrum will not be enough to on their own to cope with the continued rise in data traffic. Building more cells alone is not a solution, and it will be necessary to address costs and pricing

* The challenge needs to be approached both from the network, through policy-based control including tiering and maybe traffic-shaping, backhaul optimisation, and offload through femto cells or WLAN, and from the business side with pricing, potential tiered offers and segmentation

* Techniques have to be deployed to manage traffic to deliver customer experiences, particularly for cloud and TV services

* Since no single method of addressing capacity issues provides a complete solution and therefore a combination of offload, traffic management and segmentation is recommended.

* Mobile data optimization that includes content transformation is a crucial element in increasing the efficiency of data and video transport, by reducing the over-the-air payload on the RAN, and improving the subscriber experience with faster page loads and lower monthly data usage.

The companies that go to market with 4G services will have to be able to sustain them. The networks themselves will drive huge growth in data traffic. But changing business models also have the potential to explode transaction volumes. Not surprisingly, wholesale will be an important part of the 4G mix. The wholesale models most frequently cited are bulk access, machine-to-machine, and mobile virtual network operator.

However scalability and sustainability will also affect billing systems. CSPs will need to invest in the next generation of business-support systems (BSS) to manage customer-facing operations such as product, order, customer, and revenue management. 4G involves both capital and operating expenses, and those investments will have to be made simultaneously. CSPs will need to shift their perspective from cost to revenue management. For that, they will require more sophisticated policy and charging solutions.

The market dynamics of the Web2.0 will impact the LTE business models because it will be difficult to charge the user directly for the use of Web2.0 applications (that will run fast and smooth over LTE pipes) because Internet applications are associated with free usage. In view of this Telcos could exploit more indirect revenue sources : in a 'multi -sided' market structure the telco transactional platform can facilitate improved interactions and transactions between people and organisations : between advertisers and the end users.

Businesses can capitalise on 4G LTE for a wide set of applications, some of which are purely 'horizontal' while others are highly sector-specific, addressing needs unique to the industry. LTE's advantages are of greatest relevance to applications for personal communication and collaboration, CRM and project management. LTE will deliver improvements in the performance of many existing applications, and make feasible new applications that depend on reliable high speed or responsive data transfer.

Within the next decade, and probably by 2015, one trillion devices will be connected to the network – most not phones – moving the communications industry from quad-play or multi- play to "Tera-play". And LTE is one of the key enablers of a Tera-play world. For service providers, the benefits of Tera-play could be substantial with opportunities to drive revenue from an increasing number of high-value, multi-device, multi-service customers and infinitely larger personal and community networks.

Bottom Line : LTE Monetisation is predicated on many factors that play at the same time : dynamic data pricing / billing , appropriate device porfolio , Enterprise Verticals , wholesaling bandwidth , QoE and CRM , business model innovation , Cloud , OTT partnerships...and most of all : very clever thinking !!

---♠---

3.4 : LTE Revenue Kickstart : How Top Telcos succeed

According to the GSMA, LTE users consume on average 1.5 GB of data per month, which represents almost twice the consumption of non LTE data users, many of whom exceed their data limits .Furthermore, savvy operators have been innovating, with new offers and business models, to better capitalize on the demand, speed and capacity of LTE. All of these mean more revenue opportunities but as mobile pricing strategies evolve unlimited data plans are becoming less prevalent while tiered pricing models, based on different data entitlements, are more commonplace. These pricing models require operators to put in place usage safeguards and notifications for their customers to avoid bill shock besides offering other value added functionalities.

Many LTE operators are leveraging LTE speed to differentiate, by using tiered pricing to create perceptible value based on speed and encourage customers to upgrade to higher packages. Here are some examples showing different approaches to speed tiers: Telia, the world's first LTE operator, uses speed to enhance the value of post-paid offers and attract more post-paid customers. Telia applies a speed cap of 20 Mbit/s on its 4G pre-paid offers and provides maximum speeds up to 100Mbit/s to its 4G post-paid customers. Orange France uses speed to differentiate with its sub-brand Sosh; Orange France offers maximum LTE speeds up to 150 Mbit/s on its main contracts, whilst Sosh offers speed tiers varying from 14 Mbit/s 3G speeds to 150 Mbit/s ; Swisscom's Natel Infinity plans offer the always popular unlimited data, using speed as a differentiating value . However, the speed is reduced to 64kbit/s when a specific amount of data is consumed. They stated in their 2013 report: "Figures from recent quarters shows that customers switching to Natel infinity are generating higher revenues (ARPU)"

LTE has brought the opportunity to not only use speed as a differentiator but also to upsell high-speed. Some operators have been leveraging LTE speed to develop different upsell strategies: A1 Telekom Austria is offering high speed add-ons to supplement base tiered packages with lower speeds At 1Telekom Austria was also offering its business customers a data add-on (Eur 8.25) with night time access (22:00-08:00) to unlimited data, at full speed as per their base package. This is an interesting approach to managing speed and network resources whilst offering the always popular "unlimited" to create a new data revenue stream. Another approach adopted by Ooredoo is to offer tiered data packages and reduce the customer speed when they reach their limits before the end of the month. Customers can then purchase various data add-ons "Extra packs" to restore the speed.

According to research firm Juniper, almost 75% of roamers worldwide do not use data services. Most roamers choose to turn off mobile data or substantially reduce usage for fear of bill shock. The EU Parliament has tackled this issue by voting to abolish roaming charges altogether from 15 December 2015.The question therefore for many operators today is how to keep driving revenues from travellers.T-Mobile US have implemented an interesting approach. They offer unlimited data whilst roaming as part of their "Simple Choice Plan" ; however the speed is limited to 128Kbit/s. Whenever customers require higher speed, they can purchase high speed data roaming passes that provide total cost control; they offer 1 day/100MB, 1 week/200MB and 2 weeks/500MB passes. The plans are also subject to fair usage rules and include unlimited SMS while roaming.

This approach is very attractive for customers as they can access data abroad for free even though at low speed; yet the operator can still generate data roaming revenues by upselling high speed roaming passes.Real-time usage notification can prevent bill shock by empowering customers to take control of their data usage. Operators can configure different thresholds (e.g. 50% and 80%) to trigger notifications which can create upsell opportunities for those who have reached their data usage allowance. Bill Shock push notifications are superior to SMS and allow the use of graphically rich formats that better reflect the operators brand and product marketing requirements.

In a recent survey (Telecoms Intelligence BSS), 84% of operators revealed that they will invest in solutions for smart upsell offers, triggered by real-time context information, such as network usage, application access, location and more). This will help operators to maximize upsell opportunities and sales conversion rates as the offers are more relevant and timely.For example, a subscriber on a low speed data package, accessing Netflix, could trigger a high speed add-on offer; Another subscriber trying to access Facebook with no data package could trigger a data pass offer with unlimited access to Facebook. One operator successfully used this approach to stimulate data adoption by upselling real-time data passes (10MB for 1 day) to subscribers who had no data plan when they tried to consume any data. They used data passes to provide total cost control and remove any fear of bill shock .

In today's multi-device ownership market, allowing multiple users and mobile devices to share a common pool of data is proving a win-win for operators and customers . With shared data, operators are leveraging LTE capacity to extend their user base, accelerate data usage and spend. It is currently one of the most popular plans among LTE operators worldwide. Operators who have implemented shared data plans include AT&T, Verizon, T-Mobile US, EE, Orange Slovakia, Telstra, Telia and more.

Shared data is proving to be a real success story for mobile operators. As an example, AT&T's mobile share accounts tripled year on year to reach 11.3 million accounts and represent 45% of the post-paid subscriber base, with an average of 3 devices per account. AT&T also reported a high 46% take-up rate of their larger data plans with over 10GB of data; this represents a 70% growth from the previous quarter.

AT&T has further extended the concept of shared data to cars with Audi: "Audi's AT&T-powered in-car LTE service starts at $99 for 6 months, 5 GB". "Additionally, starting this summer Audi will become the first car maker to offer an AT&T Mobile Share shared data plan option to AT&T customers who would like to add their vehicle to their existing Mobile Share plan".

Content is increasingly being used by mobile operators to augment the value of their offerings, attract customers and stimulate data usage. LTE reinforces this trend enabling operators to offer high value, even high definition mobile content and enhance the customer experience. More operators are partnering with OTT and content providers to package various types of high value content on LTE. These include music streaming with Spotify and Deezer, TV, films, social media such as Facebook and gaming.

Imposing dynamic video management solutions can reduce the overall video traffic by 40-50% thereby significantly reducing congestion issues. For example, Vodafone UK launched LTE with Spotify premium or Sky Sports included in its Red 4G plans. They are also partnering with Netflix to enable customers to enjoy TV programmes and films on the go for 6 months, as part of their plans "as of July 2014, anyone signing up or upgrading to a Vodafone Red 4G plan can opt for Netflix as their entertainment pack of choice".

T-Mobile Netherlands sells the Deezer music streaming service to its customers, enabling them to pay directly via their mobile accounts, which is of benefit to subscribers as it's simpler; it also enables Deezer to benefit from a new sales channel and reach the operator's subscriber base indirectly. Offering high value content has proven to significantly reduce churn. Orange France for instance has indicated that "customers with an active Deezer connection are twice less likely to terminate their Origami offer".

Application service passes offer a data entitlement to use a specified application or group of applications for a defined time period. Application service passes provide total cost control and simplicity, which removes any fear of bill shock. This, combined with the popularity of some applications and social media, enables operators to stimulate data adoption and usage. A number of operators have launched application service passes. For example, Axis Indonesia offers daily, weekly and monthly Viber only passes. Movistar Mexico also offers 30, 7 and 3 days data passes with unlimited access to a bundle of popular apps such as

Facebook, Twitter, WhatsApps and e-mails, in addition to some data allowance. However the speed is reduced to 64Kbps after 1GB of unlimited app consumption.

By providing real-time visibility and control to customers, operators can stimulate trust, customer satisfaction and subsequent loyalty. This means real time notifications for all customers whether pre-paid or post-paid and self-care dashboards on the device . Real-time notifications also represent an opportunity for operators to upsell, as LTE subscribers regularly exceed their data limits. EE for example sends notifications at 80% and 100% of data usage; the notification includes a link for customers to buy a one-off data pass. Customers must then either buy a data pass or their data access is blocked until the next bill.

Another operator revealed that they achieved, on average, over 20% of sales conversion simply through up-selling data with threshold alerts. Operators can further give customers the flexibility to "buy now" on the device and have their service delivered immediately anywhere; this approach is viewed by 74% of operators as the most successful strategy to build loyalty and increase ARPU; this can also generate reduction in customer service and sales costs.

In a high speed data market, real time customer experience management would only be possible by a robust BSS.The BSS platforms need to be more of a "component" based architecture that allows for interchangeability, is robust, and scalable. A component based architecture creates flexibility for changing out solutions to keep pace with technology advancements and competitive pressures.Today's data use cases such as data service passes; application service passes and shared data plans are difficult and costly to implement on traditional, inflexible IN platforms.

New dynamic pricing models are also emerging where users can purchase additional entitlements and have them charged and provisioned in real-time. Shared Data is a use case of policy and charging control (PCC) that can be deployed in either policy or charging environments. Policy Manager (PCRF) solutions can be used to control bandwidth and consumption to minimize congestion in the face of explosive video consumption as well as allowing a flexible and contextual billing regime.

---◆---

3.5 : LTE Business Case : Impact of the ATCA Blade

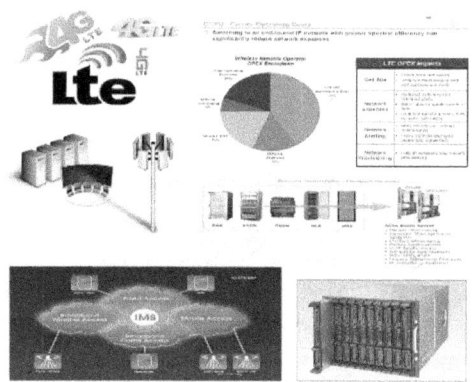 Today's service providers face three critical challenges: getting to market quickly with high-value, converged multimedia services, optimizing infrastructure costs and delivering true carrier-grade performance. To best serve their subscribers and drive new revenues, service providers are transforming their intelligent networks into intuitive networks — that are device-aware, application-aware and access-aware. New differentiating service bundles are being prepared with attractive applications, such as rich voice, gaming, presence, instant messaging, video conferencing and sharing and personalized mobility.

The build out of today's 4G networks such as LTE requires a dramatic increase in computational resources to adequately deliver flexible telecommunications services to mobile subscribers. Yet business conditions also necessitate that new markets are approached incrementally. The challenge for telecom carriers is to reduce the cost of serving the first subscriber in small or cost-sensitive markets. The primary challenge in serving small LTE subscriber bases is that traditional core network architectures require high capital expenditures just to serve the first subscriber.

Networks, whether entry-level or full-scale, are traditionally built using separate network elements for each of several different functions. And most network elements have been deployed with a pair of carrier-grade servers to achieve redundancy with an active and a standby configuration. Thus, a new network with 10 network elements requires 20 servers just to provide service to the first subscriber.

Furthermore, because the network is designed to eventually support a large population of subscribers, the servers would remain underutilized until the subscriber base grows to the expected population. The ROI for small and emerging markets has therefore been limited by these high capital outlays. High operating costs for maintaining the servers and providing data center floor space, power, and cooling have also hindered new service opportunities.

The greatest opportunity for revenue growth for wireless broadband presents itself in the form of smaller markets with less than 50,000 subscribers, thereby lowering the cost dramatically to serve the first subscriber and the breakeven point in the Operator's business case . By dramatically lowering the cost to serve the first subscriber, new networks can be built on a campus or targeted community basis with new services tailored to the specific needs of these smaller, targeted markets.

In telecommunication parlance, a "carrier grade" or "carrier class" refers to a system, or a hardware or software component that is extremely reliable, well tested and proven in its capabilities. Carrier grade systems are tested and engineered to meet or exceed "five nines" high availability standards, and provide very fast fault recovery through redundancy (normally less than 50 milliseconds). A rule of thumb is to achieve an availability of five-nines: the system is available 99.999% of the time. This equates to a stringent downtime of 6 seconds in a week or 5 minutes 15 seconds in a year.

System availability is dependent on the availability of its components. A chain is only as strong as its weakest link. Thus, if a system needs five-nines availability, then the software should provide six-nines availability and the hardware should provide six-nines availability ($0.999999 \times 0.999999 = 0.999998$).

Scalability is often in reference to architecture. A system that has five units can scale to fifty easily because the architecture allows for it. On the other hand, a system designed specifically for five units cannot scale to fifty because its architecture is inadequate. A modular hardware architecture and decoupled software architecture enable you to deploy IMS services on a very small scale (a single node) or very large scale (a multi-node, high-capacity system). Various dimensions that govern system capacity (such as provisioning, database, transactions and signaling) can all scale separately, so you can apply investments very efficiently.

The global Advanced Telecommunications Computing Architecture (ATCA) standard incorporates the latest advancements in high-speed interconnect technologies, next-generation processors and platform management capabilities. Computer equipment built to ATCA standards will work effectively in the network core of a wireline, wireless or cable provider. Service providers reap the benefits in faster time to market, lower costs and accelerated pace of innovation to introduce new features and services.A typical Service Engine forms the core of platforms built on the ATCA standard to incorporate hardware redundancy, a fault-tolerant software architecture and self-monitoring/ self-healing features.

There are separate cooling zones for redundant components, separation of switch hubs to prevent accidental removal or damage, and enhanced fault detection and handling. The software used on ATCA products is enhanced to incorporate improved reliability mechanisms such as self-stabilizing and fault tolerance. Self-stabilizing software means that the system will more readily converge to an error-free state autonomously. This can be achieved through higher coverage of hardware and software faults, an approach that is derived from Failure Mode, Effects, and Criticality Analysis (FMECA) military standards.

The message for service providers is clear : selected IP platforms are ready to deliver the reliability and availability necessary for real-time, multimedia-rich content, including voice. In an all IP world , multi-core processors coupled with powerful virtualization technology enables the consolidation of all the physically discrete carrier-grade servers into a very attractive platform for low-end scalability. Replacing 20+ carrier-grade servers with either 2 blades or 2 carrier-grade servers based on multi-core processors represents a dramatic way to lower the cost of the core network elements required to serve the first subscriber; this type of radical consolidation represents at least a 10:1 reduction in initial CapEx, plus a comparable reduction in recurring operating expenses.

Forward thinking Telcos must capitalize on the advantages of IP for converging voice, data, and multimedia services on a single unified, cost-effective core infrastructure running on ATCA carrier grade blades. The maturity of IP standards and quality of service (QoS) on IP networks opens up new possibilities for carrier applications. Converging voice and data services over a single IP backbone (such as LTE) maximizes network efficiency, streamlines the network architecture, reduces capital and operating costs, and opens up new service opportunities.

---------------------------------------♠---------------------------------------

3.6 : Data Pricing : Fresh Insights from the Airline Industry

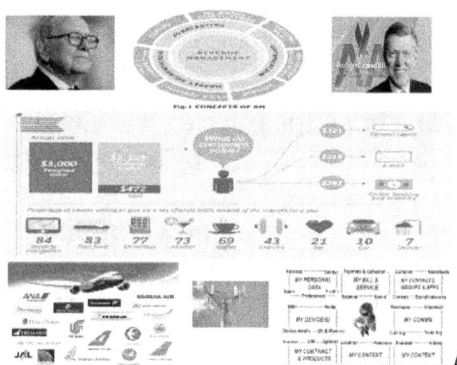

According to Warren Buffet "The single most important decision in evaluating a business is pricing power..If you've got the power to raise prices without losing business to a competitor, you've got a very good business. And if you have to have a prayer session before raising the price by 10 percent, then you've got a terrible business." If this is true then it would make Telcos in the 4G world a lousy business since Telco Execs in Africa (especially) have a prayer session on how to cut prices by + 25 % every 3 months and crow about it in the Media like pontificating politicians seeking votes during a presedential election.

The Boston Consulting Group Report (The Internet Economy in the G-20: The $4.2 Trillion Opportunity,) surveyed about 1,000 people in each of several G-20 nations on what "lifestyle habit" they would give up instead of the Internet for a year, including sex, alcohol, showers and cars. Most of the results for items like coffee, chocolate and fast food were steady with averages of 70-80 percent. Japan topped the list of citizens who would make the sacrifice, with 56 percent who would abstain from sex. Brazilians were the least likely to give up sex for the web access – only 12 percent surveyed would give it up. American and South Africans were most attached to their vehicles – only 10 percent each were willing to give up their cars for the Internet.

Another interesting finding was the perceived value of the Internet versus its actual cost. For instance, Americans value the Internet at $3,000. According to BCG, it's value is actually $472 – an incredible markup in price based on perception.

So in light of the above : is there any hope for Telco execs who are intent on destroying the profitability of the Industry by engaging in vicious price battles that threatens the sustainability of the Industry ???? Thats where Yield Management can provide some insights on how best to price data. Robert Crandall, former Chairman and CEO of American Airlines, gave Yield Management its name and has called it "the single most important technical development in transportation management since we entered deregulation." Ditto for Telco De Regulation

Yield management is the process of understanding, anticipating and influencing consumer behavior in order to maximize yield or profits from a fixed, perishable resource such as airline seats or hotel room reservations or advertising inventory or Internet bandwidth . Its core concept is to provide the right service to the right customer at the right time for the right price.This process can result in price discrimination, where a firm charges customers consuming otherwise identical goods or services a different price for doing so.

Believe it or not the telco and airline industries have much in common in terms of perishable inventory , large sunk low marginal costs and varying predictable demand volume. Yield management principles would indicate that, as the costs of using network capacity are minimal, it makes sense to use as much of it as possible when it is available and maximize revenue from it when it is in shorter supply. Unused bandwidth is lost forever.

It is surprising if not shocking that Operators are utilizing on average only 35 to 40 percent of their network capacity. It takes some creativity to turn this huge dormant asset into profits.There are several scenarios in which yield management can be used to increase revenues.

• Improving Market Segmentation and Pricing : Market segmentation enables enterprises to cater products and services, including pricing, targeted at the buyers in each segment. The basic idea, which makes segmentation an effective tool to increase an enterprise's bottom line, is to charge more for products targeted at customer segments with a higher willingness to pay.

• Monetizing Unused Bandwidth : Telcos stand to increase profits through the creative monetization of their unused network bandwidth. Similar to the empty airline seat, network bandwidth is a perishable resource with a very low marginal cost, so filling the underutilized bandwidth with revenue-producing network traffic will have a direct, positive impact on the bottom line.

• Dynamic Pricing : Market segmentation and associated segment pricing aims to maximize profits by fixing prices at levels that are optimal for targeted segments. Dynamic pricing encompasses adjusting the prices to changing market conditions and/or the status of the Telco's resources.

• Increasing Profits at Peak Utilization : Although dynamic price adjustments can be used as an effective mechanism to off-load the network during peak usage, periods of peak utilization by definition are synonymous with periods of peak market demand and, as such, should be assessed for opportunities to increase revenues.

• Reservations : The mobile network's bandwidth management functions enable Telcos to reserve bandwidth for specific customers in advance of their actual bandwidth usage. When a customer's reservation is in effect, the network ensures that the customer receives the requested bandwidth even if there are other customers competing for the same bandwidth.

• Pricing flexibility : Real-time charging functionality provides Telcos with pricing flexibility at least on par with, if not better than that used by the airline industry. Telcos can charge based on the attributes of provided services, customer characteristics, context, network state, historical usage, etc.

Besides increasing revenue Yield management can also reduce the need to increase capacity, resulting in savings in investment for providers, which can be passed on to consumers as lower costs. The value of yield management for mobile broadband is an opportunity for service providers to manage the quality of a user's experience while achieving increased revenues in the context of the exponential CAPEX costs associated with servicing the global demand for mobile broadband services.

3.7 : Telco CRM : using Social Media as a strategic weapon

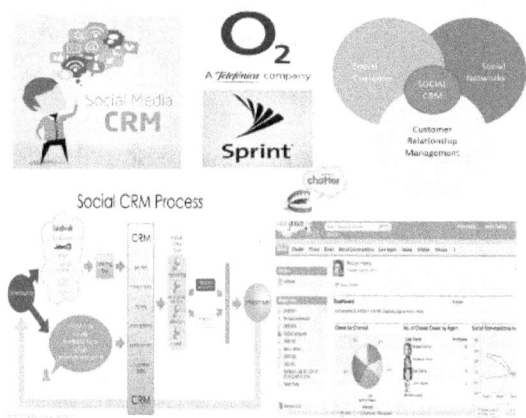

Unless you live in a cave or on Mars, you should know that social is the way to go. Social media – blogging, online social networking, and micro-blogging – have become so pervasive that it is almost unthinkable for a business entity – at least those who want to remain relevant !! In telecom, social media have transformed not only business models but the very concept of customer service. Emerging markets have embraced social media with gusto. Both India and Brazil represent some of the most aggressive growth, where more than 90 percent of online survey respondents report having an account on a social networking site. The reasons for this social media explosion in the emerging markets can be attributed to the concentration of Generation Y and younger, the cultural emphasis on maintaining regular contact with friends and family, and the influx of mobile technologies.

The IBM Institute for Business Value surveyed more than 1,000 consumers worldwide to understand who is using social media, what sites they frequent and what drives them to engage with companies. What the results showed may come as a surprise to those companies that assume consumers are seeking them out to feel connected to their brand. In fact, consumers are far more interested in obtaining tangible value, suggesting businesses may be confusing their own desire for customer intimacy with consumers' motivations for engaging. For companies that have been taking a "build it and they will come" approach to social media, these consumer findings are a wake-up call that much more needs to be done if they want to attract more than the most devoted brand advocates.

Telefónica Europe wanted to develop a European-wide social media strategy to align social media use across each of the company's business units and to help staff use it to enhance the whole customer experience. Telefónica Europe started by evaluating the company's current use of social media across all O2 branded businesses. Not only did this help to determine the strategic aims of each business unit, but, more importantly, it helped to highlight where the business was currently realizing value from social media, and the quick wins and opportunities for improving customer experience in the long run.

Once the audit was complete, the next step was to develop a consistent strategy for the use of social media across multiple European markets in support of Telefónica's commercial brand, O2, and its Brussels-based Public Affairs department. This included developing three-year aims and objectives for Telefónica Europe's social business strategy; the whole process also took into account each local market's needs and conditions.

Salesforce.com was selected as the platform which allows a fast delivery of the solution in the cloud and also provides an integrated module for social media communication called "Chatter" lifting Web 2.0 Technology into the Cloud. Due to the scalability and multi-tenancy of the Cloud solution, further CRM applications can be easily integrated.Telefonica are developing, testing and embedding social media programmes across Europe in different departments and teams including Customer Experience, Brand, Communications and Customer Contact Centres. Education and training for 29,000 employees was implemented to embed the social media strategy and standardised processes across the business to ensure best practice and maximum return from social media.

As a leading Telco with over 23 million customers, O2 (subsidiary of Telefonica) decided to embrace social channels, as it is increasingly seeing a change in customers' buying and service expectations—with a growing preference to use the online channel. O2 started a business transformation journey to a multi-product, multi-channel company whilst continuing profitable growth. Evidence of this transformation can be seen in some of the services such as the new online shopping experience for small office/home office businesses. A Chatter app clears the line for employee communication O2's in-store support staff—called "gurus"— to collaborate on issues and help customers immediately, often while they are still standing at the counter.

In the Telco industry, loyal customers are the key to success. Nobody knows that better than Sprint, rated #1 for customer satisfaction and the third-largest telecommunications provider in the United States. When Sprint wanted to use new social media platforms to share information across groups and manage relationships with business customers more efficiently they consolidated customer information, automated processes, and built apps to make it easy to share data with retail stores. Information on business customers from multiple CRM systems was consolidated into customer profiles in the social media platform : a one-stop-shop for our sales teams for business processes and customer intelligence.

More than 6,000 employees now use a Cloud platform to track accounts, contacts, and opportunities for improved visibility and real-time analytics.Sales Reps can quickly identify the best opportunities driving new leads and reducing the time to close enterprise deals by 25%. And, a fully-integrated configure-price-quote system helps Sprint's sales teams quickly build complex pricing models using up-to-date account information.

Customer visibility is enhanced by a Chatter application, which helps reps share and information and collaborate on deals so they can address customer needs quickly and provide consistently great service. Using a custom Internet portal, retail staff can easily share business leads with sales, and no leads get dropped. In the future, retail personnel will be able to collaborate with customer service professionals to solve customer issues on the spot. Custom apps built on the social media platform automate processes including managing waitlists for hot new phones, scheduling corporate briefings, or tracking churn data, so reps can proactively reach out to customers in danger of defecting or counteract competitive promotions. Another app manages the discount approvals process and recaptures almost $70 million in unauthorized discounts each month.

The benefits are social media based CRM are deep provided it is implemented strategically. First, there is the social interaction itself, which can provide direct value to the business through revenue from social commerce and cost savings when used for customer care or research, for example. Plus, social networking enables rapid, viral distribution of offers and content that may reach beyond what could be done in traditional channels – all with endorsement from connections people trust. But that is just the beginning.

Companies also can use social platforms to mine data for brand monitoring and valuable customer insights, which can spark innovations for improved services, products and customer experiences. In a constant cycle of listen-analyze-engage evolve, Telcos can optimize their social media programs to continually enhance their business. But :Telcos beware. Hell hath no fury like a customer scorned on social media !!

--♠--

Chapter 4 : Network in the Cloud

4.1 : Cloud RAN (C – RAN) Reloaded : NFV + Small Cells + LTE A

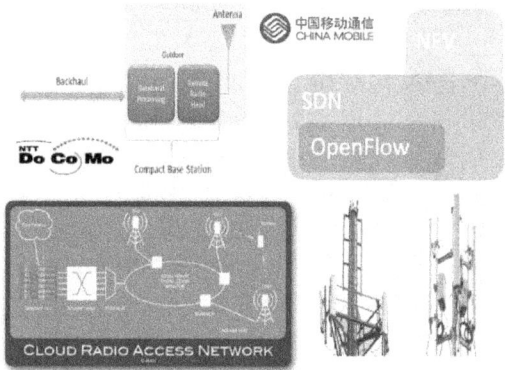

As all Telco engineers know that in a typical mobile deployment, each base station serves all the mobile devices within its reach. Each base station has its digital component manage its radio resources, handoff, data encryption and decryption and an RF component which transforms the digital information into analog RF. The RF elements are connected to a passive antenna that transmits the signals to the air. Each base station should be placed in the geographical center of its coverage area.

But even when such locations are selected, the mobile operators may have difficulty in renting the real estate, finding proper powering options, securing the location and protecting the equipment from weather conditions. Those cell sites carry with them a continuous stream of OPEX to address the high rental rates for real estate, electrical expenses, cost of backhaul for the cell site and security measures to protect the location from intruders.

Enter the latest architectural paradigm : C RAN !!! The basic premise of Cloud RAN is to change the traditional RAN architecture so that it can take advantage of technologies like cloud computing, Software-Defined Network (SDN) approaches, and advanced remote antenna/radio head techniques.C-RAN architecture is not bound to a single RAN air interface technology. In essence, conventional terrestrial cell site base stations are replaced with remote clusters of centralized virtual base stations which can support up to a hundred remote radio / antenna units.

This is achieved by centralizing RAN functionality into a shared resource pool or "cloud" (the digital unit – DU, or baseband unit – BBU) which is then connected via fibre to advanced remote radio heads ("Radio Units" – RU) sited in different geographical locations in order to provide full coverage of an area. The radical concept can even use banks of x86 servers to connect cellular calls rather than traditional wireless base stations.

From a business perspective,C-RAN will deliver significant reductions in Opex and Capex due to reduced upgrading costs. A major reason for this is the aggregation and pooling of the DU computing power which can be assigned specifically where needed e.g. the load situation over time and space for indoor/outdoor cells, am/pm hours, weekday/weekend, and so on.

As a result, single cells do not need to be dimensioned for peak hour demands, but rather the processing power can be pooled and assigned on an on-demand basis. The processing power savings achieved should also leave processing headroom for any further potential technology enhancements (e.g., LTE-A features) without the need for further CAPEX. C-RAN skips the need for a high-bandwidth, low latency (X2), synchronized interface between the geographically distributed base station because the computing resources of the multiple transmission points' BBUs are all located within the same hardware.

C-RAN slashes capex because fewer BBUs are needed, which reduces opex because fewer BBUs means less energy consumption and diminished maintenance costs. The reduced energy consumption makes C-RAN a "green" alternative, with China Mobile estimating 71 percent power savings vs. traditional RANs.

Furthermore, interference management will also benefit from C-RAN network architecture as technologies like dynamic eICIC schemes will be enabled, especially in a HetNet deployment. Heterogeneous networks will require small cells to be independent, intelligent and ubiquitous to avoid the cross-interference mayhem, yet be in synch and orchestrated with macro cells (including Cloud – RAN topology).Small cells are poised to become the most commonly used node for cellular access in the next-generation HetNet.

C RANs will likely take their place beside traditional base stations and emerging small-cell base stations as another tool for building cellular nets.The success of many new 4G network deployments will depend on the use of outdoor and indoor small cells to extend coverage and increase capacity in areas poorly served by macrocell networks. Operators are also considering proposals to deploy more efficient CloudRAN architectures requiring high speed CIPRI front haul links between remote radio heads and pools of baseband units.

According to Maravedis Cloud-RAN economics only be realized by harnessing standards to ensure interoperability and reduce cost. That, in turn, will create a whole new ecosystem, and operators must resist any attempts by their suppliers to hijack standards for software-defined networking or cell site equipment. Otherwise, this fledgling architecture will remain confined to a few pioneers with the resources to build their own ecosystems, like China Mobile.

China Mobile, the world's largest carrier with 700 million subscribers, has been spearheading trials and plans to deploy systems as early as 2015. Japan's NTT Docomo said it will follow in 2016, and a third unnamed carrier is now preparing plans for C-RANs. China Mobile aims to lower the cost of C-RANs to less than $30 per LTE sector, down from about $10,000 two years ago. It will start a second round of trials later this year using servers equipped with PCI Express cards to handle baseband processing. Each card will pack four FPGAs using silicon cores, each FPGA capable of handling 12 LTE sectors.

As MNOs face rising CAPEX bills to meet mobile data demand combined with falling ARPU, they must explore radical new network designs. With Cloud-RAN, they can virtualize baseband processing functions for hundreds of sites on a server or base station hotel. By consolidating individual Base-station processing into a single or regional server farm Investments on Cloud Radio Access Network (RAN) Infrastructure are expected to exceed $6 Billion by 2020, according to a new report from SNS Research. Distributed antenna technologies (DAS) will get a new lease on life, supporting coverage extension for C-RAN sites.

This sector will open up $1.3bn in new revenues for antenna providers. Pure C-RAN faces many barriers, such as over-reliance on fiber to link sites and basebands and immature standards, but most operators will inch towards C-RAN using hybrid models. Development of microwave fronthaul technologies will be critical to improve the C-RAN business model .

Whatever the challenges C-RAN offers a revolutionary approach to next-generation cellular networks deployment, management and performance. Fiber, needed for fronthaul, is crucial to C-RAN deployment, so it is no wonder that fronthaul is constantly brought up as Cloud RAN's biggest challenge. Fronthaul connects RRHs to the aggregated BBUs, with traffic then backhauled from the BBUs to the IP core or evolved packet core (EPC).

NTT DOCOMO, Japan's leading mobile operator and provider of integrated services centered on mobility, announced today it will begin developing high-capacity base stations built with advanced C-RAN architecture for DOCOMO's coming next-generation LTE-Advanced (LTE-A) mobile system. The new architecture will enable quick, efficient deployment of base stations, especially in high-traffic areas such as train stations and large commercial facilities, for significantly improved data capacity and throughput.

Advanced C-RAN architecture, a brand new concept proposed by DOCOMO, will enable small "add-on" cells for localized coverage to cooperate with macro cells that provide wider area coverage. This will be achieved with carrier aggregation technology, one of the main LTE-Advanced technologies standardized by the Third Generation Partnership Project (3GPP). The small add-on cells will significantly increase throughput and system capacity while maintaining mobility performance provided by the macro cell.

For NTT DoCoMo high-capacity base stations utilizing advanced C-RAN architecture will serve as master base stations both for multiple macro cells covering broad areas and for add-on cells in smaller, high-traffic areas. The base stations will accommodate up to 48 macro and add-on cells at launch and even more later. Carrier aggregation will be supported for cells served by the same base station, enabling the flexible deployment of add-on cells. In addition, maximum downlink throughput will be extendible to 3Gbps, as specified by 3GPP standards.

C-RAN is typically thought of as a large-scale urban macro solution, but the concept of pooled baseband serving n number of radio access nodes can apply to a variety of scenarios, such as small cell underlays (using micro RRUs), so-called Super Cells, and outdoor/indoor hotzone systems. These models, identified and defined partly through the NGNM Alliance, could prove an attractive way to introduce and develop C-RAN technology. Given the traditional RAN's coverage restrictions and limitations of transmission and reception signal support, the benefits of deploying a C-RAN infrastructure are clear.

Bottom Line : The C-RAN, as a centralized, general purpose processing solution, enables the efficient use of network resources. Based on open-platform and base station virtualization, C-RAN provides an ideal architecture for LTE-A functionality as well as being complementary to next-generation SDN and NFV deployments. Many major mobile operators across the globe are preparing to incorporate the cloud into their existing RAN platforms. We anticipate that 2014 will move the C-RAN beyond the "cloud hype" as operators gain a competitive edge through integrating the C-RAN in their LTE-A migration.

---♠---

4.2 : Telco Clouds : All aboard the OpenStack express

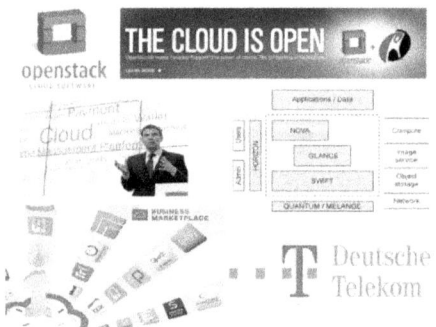

For all you Telco Cloud builders something is floating up there : they call it Openstack !! OpenStack is a free and open-source software cloud computing platform that is primarily deployed as an infrastructure as a service (IaaS) solution. The technology consists of a series of interrelated projects that control pools of processing, storage, and networking resources throughout a data center, able to be managed or provisioned through a web-based dashboard, command-line tools, or a RESTful API. It is released under the terms of the Apache License. Founded by Rackspace Hosting and NASA, OpenStack has grown to be a global software community of developers collaborating on a standard and massively scalable open source cloud operating system.

So who is the big user of Openstack besides NASA? Well how about the Large Hadron Collider (LHC) project , which generates 1PB of data every second ?? CERN started using the OpenStack private cloud back in 2011 in the testing environment, upgrading more recently to the fifth version of OpenStack : the Essex release. Moving to a mammoth-scale infrastructure-as-a-service (IaaS) cloud based on OpenStack has helped the European Organisation for Nuclear Research (CERN) significantly expand its compute resources and support more than 10,000 scientists worldwide using the infrastructure to find answers to questions such as what the universe is made of. The big vision for CERN's private cloud infrastructure is to be able to scale up to hosting 15,000 hypervisors on the cloud by 2015, running between 100,000 and 300,000 virtual machines !!

A major strength of OpenStack is its ability to more easily enable hybrid cloud platforms – the combination of public and cloud operations that appeal to larger enterprises looking to combine the savings of commodity public clouds for some operations with the security and control of private clouds for other apps. Now that VMware has joined OpenStack consortium the plattform is expected to gain even more momentum as well among major telecom players operating their cloud subsidiaries. Today's Telcos are struggling to support ever-expanding demands from both consumers and enterprises, including the need to transmit and store increasing amounts of data. As the Internet of Things grows, these data throughputs will only rise alongside user requirements.

According to pundits while other Cloud solutions can provide adequate support for boosted data traffic levels, OpenStack does so in a cost-efficient and elegant manner that other technologies just can't match. Competitive solutions are three to five times more expensive than OpenStack deployments . OpenStack's unique features and functionalities , such as allowing service providers to grow or shrink their offerings to match service peaks and no licensing fee , make it an ideal Telco Cloud platform.

DT is one company that has made the Open Source Cloud as the centrepiece of their global services strategy. Deutsche Telekom offers a portfolio of over 30 different cloud services that encompass infrastructure, developer environments, collaboration, business applications and security as a service. By employing the cloud service brokerage strategy, Deutsche Telekom has become the cloud partner that its millions of business customers are looking for. The loyalty and lifespan of those customers will be dramatically increased as they receive innovative, business-critical services from a provider they already know and trust.

Deutsche Telekom enhances its set of cloud technologies with OpenStack. The open source cloud operating system makes it easy for software partners (ISVs) to integrate their cloud applications in the Deutsche Telekom infrastructure and its new Business Marketplace, removing technical obstacles. Deutsche Telekom's Business Marketplace is an online platform which will offer cloud services for small and medium businesses started in 2012. Their goal was to offer our customers a rich set of business applications out of our cloud. Most mobile cloud applications are sophisticated mash-ups of all sorts, including applications that incorporate core mobile phone and Telco capabilities, like location, presence, phone calendar, address book, and cameras.

The Business Marketplace consolidates innovative business customer solutions from DT partners. Usability is a particular focus: business customers can find, book and manage the cloud-based applications with a single click. In addition to detailed information on all products, users can also see other customers' ratings and try out the products free of charge before buying them.

Business Marketplace also lets companies keep track of their application licenses and employees' access privileges at all times, as well as view their billing data. The users receive a single bill for all ordered applications on a monthly basis. One of the most valuable services, that can enfranchise many thousands of new ASPs, is the 'Bill on Behalf-of' (BoBo) ability for them to be paid through their customers' phone accounts.

Comprehensive payments and settlement and the associated business infrastructure, is a critical component of Mobile Cloud Computing and DT are leveraging this element . The operator also plans to contribute to the development of OpenStack and Deutsche Telekom has created a growing team of in house engineers that will work to harden and secure OpenStack. They have already started offering a security package for SMEs to protect them from viruses and attacks from the internet.

DT's aspiration is for €2bn in additional revenue by 2015 built on a track record of being among the first to bring leading-edge cloud innovation to customers in Europe covering domains like end user computing, enterprise networking and data center. DT firmly believes in developing an open ecosystem. It is not that the telco is afraid to innovate on its own behalf; indeed, it develops many of its applications and platforms in-house. It has, however, become very open to working with partners if they can either speed up time to market or add brand credibility.

Despite its many virtues OpenStack offers risks and rewards to telcos. On the upside, OpenStack allows telcos to more rapidly roll out cloud services at low cost.On the downside, OpenStack standards could allow end-customers to more easily migrate their cloud applications from one OpenStack telco to the next. Telco Cloud service strategies vary : some telcos are simply reselling Office 365 while adding some value-added services. Office 365 resellers include Bell Canada, KCN of the Netherlands, France-Telecom Orange, Telefonica, TeleiSonera, Telmex, Telstra and

Vodaphone etc. Other telcos see OpenStack as a potential way to battle Amazon, Google and other cloud providers. Verizon's buyout of Terremark and CloudSwitch essentially positions Verizon against OpenStack-focused service providers. Its the wild west in the clouds !!

Even the network equipment vendors are clambering on the OpenStack bandwagon . Ericsson is now previewing a tweaked version of OpenStack that will run on its network iron. Ericsson is giving the same server consolidation pitch that has been common in the mainframe market for three decades, in Unix for a decade and a half, and for the past decade or so in the x86 server racket, to push OpenStack into its telco gear. At the heart of the Ericsson Cloud System is what the company calls the Cloud Execution Environment, which runs on x86 iron, of course. It uses the KVM server virtualization hypervisor championed by Red Hat to dice and slice virtual server instances on top of physical server blades and puts KVM and the workloads that run on its virtual machines under the control of OpenStack. And if Ericsson is on board then the Chinese vendors are not far behind !!

Bottom Line : The fact that OpenStack was developed by a group of people who called themselves "open source" rather than a group that named themselves after a rain forest is probably irrelevant. The important thing is that, like Linux, it works robustly and is cheap and was built by people who knew what they were doing. The defining features of OpenStack is scalability and adaptability. As enterprises push for more open APIs into cloud platforms, both to ease the complexity of moving to the cloud and to prevent the dreaded vendor "lock-in," the pressure will mount on cloud operators of all kinds to embrace OpenStack.

---♠--

4.3 : NFV and SDN Reloaded : Get started Telcos

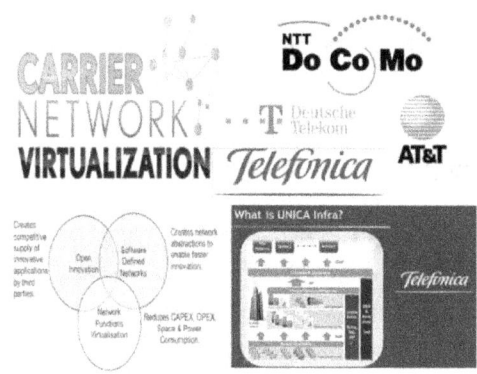

According to Deutche Telecom CEO communication networks are facing a lack of scalable and sustainable architecture to meet the challenges ahead in terms of data traffic increases, video uploads and downloads, and enhanced M2M communication. But employing software-defined networking (SDN) techniques could help mobile carriers overcome those hurdles and attract new data-centric revenue streams.

In a nutshell, SDN delaminates the data and control planes of the network and NFV virtualizes the functional elements of the network—routers, switches, firewalls—and expresses these functions as programs that run on commercial off-the-shelf (COTS) IT hardware. While they are distinct technologies, the two work together in concert to turn the network into an infinitely programmable dynamic mesh, versus a hardware-based static map. Where SDN is the network admin gone virtual; NFV is the gear gone virtual.

Today's mobile networks are limited and built upon a best-effort design, but that means they have latency issues and cannot dedicate high bandwidth to a particular user on the fly. Network virtualisation highlights the transformational path that operators are willing to take to counter the stress that financial pressures are putting on profitability while effectively and efficiently monetising data growth and reducing vendor lock-in. This trend clearly shows that, in order to be sustainable in the near-future, operators networks will require the right amount of mobility, ultra high-speed networks, cloud computing, big data analytics and security.

Research into NFV performed by leading analysts firms confirms the development of NFV and reveals major market potential. In November, Mind Commerce estimated that the NFV market in 2014 will be worth $203 million, and will grow at 46 percent annually until 2019, when it reaches $1.3 billion. The research firm states that the chief domains targeted by early NFV deployments will be IMS services and the EPC.

Last August ABI Research predicted a similar growth curve, with a potential $6 billion market for virtual networking by 2018. A new study from ReportsnReports.com forecasts that the NFV, SDN and wireless network infrastructure market will reach $5 billion by the turn of the decade, driven by rising global wireless capital expenditures and growing demand for high-speed mobile broadband. Wireless carriers will play a critical role in the SDN value chain, and that carriers will initially focus on southbound APIs and switch fabric, SDN and virtualization that will enable IMS optimization and realization of investment, and that by 2016 carriers will focus more on northbound APIs and create full development environments.

Network virtualisation allows operators to simulate network resources through SDN and NFV technologies that decouple, run and optimise different functions of the network.The industry is evolving from proprietary equipment networks to IT-based data centre networks that employ technologies such as software-defined networking (SDN), network function virtualisation (NFV), cloud-computing and big data analytics to provide a variety of converged services to consumers. NFV is highly complementary to

SDN. Network functions can be virtualised and deployed without an SDN being required and vice-versa.

According to ETSI, early NFV deployments are already getting underway and are expected to accelerate during 2014-15.Software-defined networking (SDN) and Network Functions Visualization (NFV) will drive changes in data security investment, according to a new report from Infonetics Research. Their Data Center Security Products report noted a shift in how organizations protect digital properties, including a 44 percent rise in the sale of purpose-built virtual security appliances. They anticipate a fairly significant revenue transition from hardware appliances to virtual appliances and purpose-built security solutions that interface directly with hypervisors, with SDN controllers via APIs, or orchestration platforms.

Rather surprisingly, communications service providers (CSPs) themselves, not vendors, are driving the development of network virtualization technologies. The potential to dramatically accelerate new service delivery, lower operating costs, and eliminate vendor lock-in has CSPs salivating and network equipment vendors scrambling. Vendors who sell proprietary network gear don't exactly welcome the thought of their intellectual property being replaced by standardized software running on commodity hardware. This has pushed the timeline for SDN and NFV further out, and prompted more than a few analysts to pull the hype card.The virtualization of service and control functions in the core network has been a first step in using cloud computing technology in the telco domain.

However, for a full telco cloud implementation, virtualization needs to be complemented with a complete cloud platform and management system. This must include classical network management for legacy systems, plus virtualized network function, cloud orchestration and application management to achieve the full benefits of automated provisioning and elastic scaling of the network.

Driven by the promise of total cost of ownership reduction, wireless carriers are aggressively jumping on the NFV and SDN bandwagon, targeting integration across a multitude of areas including radio access network, mobile core, OSS/BSS, backhaul, and CPE/home environment.Telecom Italia has been among the tier 1 telcos driving the move to NFV. Along with AT&T, BT Group, Deutsche Telekom, Orange, Telefonica and Verizon, the company a couple years ago pushed network functions virtualization into the spotlight by creating an ETSI group to explore the technology. The key goals of the NFV Working Group are to reduce equipment costs and power consumption, improve time to market, enable the availability of multiple applications on a single network appliance with the multi-version and multi-tenancy capabilities, and encourage a more dynamic ecosystem through the development and use of software-only solutions.

Telefonica's UNICA platform is initially focused on virtualising signaling-related functions, including IMS (IP multimedia sub-system, DNS (domain name system), SMSC (short message service centre) and OCS (online charging system). The second phase will look at virtualising functions that carry traffic such as the core packet network. Telefonica's NFV programme is notably designed to "source different functions to different suppliers" and avoid vendor lock-ins. The company wants to design a virtualised network architecture that allows vendor interoperability.Among the many capabilities offered by UNICA is the idea of multi-tenancy (where the same basic solution works for multiple organisations) or NaaS (Network as a Service), using pre-installed templates to deploy virtualised equipment in real time and with integrated resource management.UNICA promises to offer real and permanent change for Telefónica's network transforming the company into a true Digital Telco.

Meanwhile AT&T, has introduced its vision for the company's network of the future: the 'User- Defined Network Cloud.' AT&T claims their the cloud-based architecture is "a global first at this scale." The operator also announced the group of vendors that will work on implementing this strategy. The carrier expects its revamped architecture will accelerate time-to-market for technologically advanced products and services. Integrated through AT&T's wide-area network (WAN) and using NFV and SDN, the architecture is expected to simplify and scale AT&T's network by separating hardware and software functionality, separating network control plane and forwarding planes, and improving functionality management in the software layer.

This move to software-based telco environments will not only help incumbent providers become more agile and adapt to market trends and subscriber demands more effectively, but will open up the market to new players who may not have had such deep pockets needed to develop proprietary hardware. It will allow new carriers to quickly scale and compete, as they won't have to load up on costly central office equipment to get started.

---------------------------------------♠---------------------------------------

4.4 : Deep Dive : Assessing the utility of SDN for Telcos !!

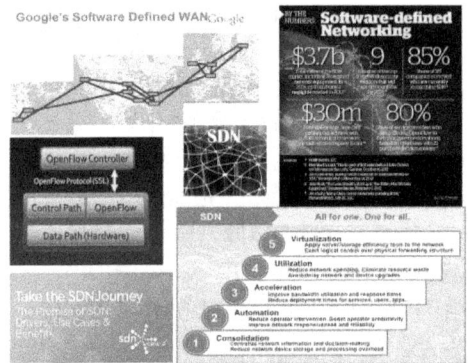 Software Defined Networking or SDN is a technological approach to designing and managing networks that has the potential to increase operator agility, lower costs, and disrupt the vendor landscape.With SDN the network becomes a programmable fabric that can be manipulated in real time to meet the needs of the applications and systems that sit on top of it. SDN promises fully automated, application-aware and adaptive adjustments to bandwidth, compute power and storage with end-user visibility is what is needed to provide ultimate QoE to the mobile user connected an IP based Telecom network such as LTE .

The root cause of a network's limitation is that it is built using switches, routers and other devices that have become overly complex because they implement an ever-increasing number of distributed protocols and use closed and proprietary interfaces. By decoupling the network control and data planes, OpenFlow-based SDN architecture abstracts the underlying infrastructure from the applications that use it, allowing the network to become as programmable and manageable at scale as the computer infrastructure that it increasingly resembles. An SDN approach fosters network virtualization, enabling IT staff to manage their servers, applications, storage, and networks with a common approach and tool set.

In a SDN, the network administrator can shape traffic from a centralized control console without having to touch individual switches. The administrator can change any network switch's rules when necessary — prioritizing, de-prioritizing or even blocking specific types of packets with a very granular level of control. This is especially helpful in a Cloud computing multi-tenant architecture because it allows the administrator to manage traffic loads in a flexible and more efficient manner. SDN allows network engineers to support a switching fabric across multi-vendor hardware and application- specific integrated circuits.

Most of the churn in mobile subscribers today is attributed to poor QoE (Quality of Experience). What's needed is an efficient and elastic system that adapts to the end-user traffic automatically and dynamically. The rapid adoption of 4G Mobile (LTE) necessitates uninterrupted availability of quality services 24/7 regardless of location or device. SDN has the capability to make this a reality. The Open Flow protocol allows the network to be programmed on a per-flow basis and thereby provides visibility at the user and application level. The capability to increase or decrease the bandwidth needed, for instance, by way of automated bandwidth signalling is one advantage. It can also adjust the number of VMs (Video Messages) and the associated storage needed proactively and dynamically with¬out any human intervention on an application basis.

Developing an SDN business involves the deployment of physical infrastructure, a network controller and a telecoms operating management system which combines operation and business support systems. The network controller is central to SDN with two main functions: virtual resource control and traffic management systems (TMS). The network controller can create a programmable, logical network that allocates resources within the physical network (access and core networks) in the most dynamic way without needing to know the actual infrastructure topology. In so doing, the operator can build the most appropriate virtual network offering multiple services.

SDN is not only an esoteric technology concept but a current reality: in 2012 Google announced that it had migrated its live data centres to a Software Defined Network using switches it designed and developed using off-the-shelf silicon and OpenFlow for the control path to a Google-designed Controller. Google claims many benefits including better utilisation of its compute power after implementing this system. Recently Japanese vendor NEC established a partnership with Portugal Telecom that will see the two firms collaborate on SDN (software defined networking) and virtualisation technology for data centers and carrier networks. The two firms claimed the agreement would enable both companies to test and assess the commercial feasibility and benefits of SDN implementation for carrier data centers , adding that SDN and network virtualisation have "exceptional potential".

According to Infonetics, telecoms plan to deploy SDNs and NFV by 2014 within data centers, between data centers, operations and management, content delivery networks (CDNs), and cloud services. In most cases, Telcos are starting small with their SDN and NFV deployments, focusing on parts of their network, in " contained domains " such as data centers, to ensure they can get the technology to work as intended.

Running in parallel Telco network architects believe that NFV (this term Network Function Virtualisation was coined by the European Telecommunications Standards Institute) will consolidate many network equipment types onto industry standard high volume servers, switches and storage, thus providing a new network production environment so as to lower cost, raises efficiency and increases agility. Network Functions Virtualisation can be implemented without the prescence of a SDN , although the two concepts and solutions can be combined to unlock greater value.

In the next decade SDN is big business. According to SDN Central (an independent market research community for SDN & NFV), the SDN market is expected to surpass $35 billion in the next 5 years.

Adoption of SDN technology has accelerated in recent years from sales of $10 million in 2007 to $252 million in 2012.The emergence of the software-defined networking market is supported by growth in venture capital investment in SDN focused companies. Venture capital funding rose from $10 million in 2007 to $454 million in 2012. It is big business !!

--♠--

4.5 : Telcos : Time to beef up your software engineering talent

In the past decade the telecommunication industry has been revolutionized by advances in three core technologies: photonics, microelectronics, and software. The emergence of high-speed optical transmission and switching plus 4G Ran is likely to fuel an already growing demand for interactive image communications, multimedia applications, and real time video services, including video conferencing, TV, and High-Definition TV. As such a deeper understanding of the architectures and protocols for broadband integrated services networks and the ability to highlight relevant performance issues becomes a critical skill.

Clearly the modern IP based Telecoms networks are more about software than hardware now. The BSS / OSS has become the brain of the network with its complex layers of middle ware to control and bill for traffic in high speed wireless data networks. In the current telecom market where the devices are smarter, the networks are accelerated and customers are well informed, the balance between OSS and BSS plays significant role in the quality of customer experience. OSS and BSS together enable the CSP's to consolidate, simplify and automate the operations.

According to IBM Tech Trends Report , mobile computing, cloud computing, social business and business analytics have gone beyond niche technology status and are now part of core IT focus. All of these technology trends require fast response times, vast stores of data, and a highly elastic backbone of networks and servers. The new software developed for clouds demands different kinds of code to take advantage of the flexibility of computing clusters.

Today's networks are facing the increasing pressures of mobility and BYOD, social media usage, and Big Data analytics. More bandwidth is required to support these trends, and IT is being challenged to reduce latency and deliver acceptable performance for cloud-based applications and services. One response to the escalating demand for faster, more efficient networks has been the emergence of software-defined networking (SDN). It's still early days for SDN, with adoption being confined largely to industry giants such as Google. But as SDN matures, it could play a critical role in helping organizations define, provision, and manage their networks.

In the US there a stampede for software talent. Companies and Universities battle to attract students with Maths skills to learn about software engineering or design exciting new platforms that leverage the Internet. Recruiters say the fiercest demand is for top-level, experienced workers with a few critical skills such as user interface, which involves designing the look and feel of a software application; mobile apps development, which entails programming for smartphones and tablets; and cloud computing software, which requires new kinds of code. Yet the Telecoms industry in MEA Region is plagued by :

• An acute shortage of people with Science , Engineering and Technological competencies combined with essential management skills
• Unavailability of readily accessible information on trends and conditions in the labor market enabling correct career and learning choices and investment decisions
• Dwindling pool of technically competent and adequately prepared candidates from the youth market to take up available jobs in the science , information , technology sectors
• Lack of funding for focused training and education projects that will ensure continuous skills upgrading to keep abreast of the technological innovation

This " penury in competence " and ensuing structural unemployment has grave implications on the competitiveness of many Telco companies . Many countries in Middle East and Africa have a high structural unemployment rate ie : the unavailability of skilled people for available jobs. Many unemployed are professional people with social sciences backgrounds that are worthless in the new InfoTech economy.

Ofcourse it is a no brainer to recruit software engineers from the " software factory " nations like India if you can afford and wish to rely on expats. However this does nothing to address the problem of structural unemployment or rising joblessness among the youth in Middle East and Africa .

Software engineers apply the principles of engineering to the design, development, maintenance, testing, and evaluation of the software and systems that make computers or anything containing software work. Some of the basic competencies a typical Software Engineer learn are :

• Study of the principles and practices of software engineering : software quality concepts, process models, software requirements analysis, design methodologies, software testing, and software maintenance. Hands-on experience building a software system using the waterfall life cycle model.

• Problem-solving and program design using C++ : Introducing a variety of programming techniques, algorithms, and basic data structures—including an introduction to object-oriented programming

• Software Testing and Quality Assurance : quality concepts, black and white box testing techniques, test coverage, test planning, levels of testing, the formation of a testing organization, testing-in-the large and special problems in object-oriented testing, documentation for testing, and inspections and walkthroughs as a vehicle for product quality

• Oriented Information Systems : Investigation of different architectural strategies for building object oriented information systems. Develop familiarity with modeling, design, and implementation techniques used in the construction of object oriented information systems.

• Software Metrics : Theoretical foundations for software metrics. Data collection. Experimental design and analysis. Software metric validation. Measuring the software development and maintenance process. Measuring software systems. Support for metrics. Statistical tools. Setting up a measurement program. Application of software measurements.

What is urgently required is an innovative industry-university-government collaboration to prepare math and science graduates for advanced study in software engineering, telecommunications, and satellite communication and provide them with convenient advanced degree programs This will ramp up the development of software and telecommunications engineering human resources, and help accelerate the development of ancillary telecoms and software engineering industries in MEA Region.

Telcos in MEA need to do more to accelerate the skilling software engineers by setting up Software Engineering Centres in the countries they operate instead of relying on universities.

---♠---

4.6 : How to monetise the mobile cloud – the Telco / CSP strategy

 For mobile communication service providers (CSPs), revenue from voice service continues to decline as a percentage of overall revenue, while global mobile data use continues its inexorable rise. Indeed, industry analyst firm Informa Telecoms & Media has predicted that revenues from mobile data will have grown from $210bn in 2009 to more than $450bn by 2015.

Cloud offers a unique opportunity to service providers that want to offer value added services like voice, video and collaboration on cloud platforms, but success will come only with simplicity and a recognition that the economics of the cloud are very different than traditional telco models. In essence, recognition that cloud models have utilization patterns like airlines, which means they are capital intensive, meaning supply and demand differentiation will be critical to maximizing yields.

Most mobile cloud applications will be sophisticated — mash-ups of all sorts, including applications that incorporate core mobile phone and Telco capabilities, like location, presence,phone calendar, address book, and cameras.In addition to the services accessed by "traditional" means such as via mobile handsets or web browsers, there is huge potential for machine-to-machine (M2M) services availing of Mobile Cloud Computing.

As mobile operators contemplate their mobile cloud strategy, it is critical that they reinforce one of the key value propositions that truly separates them from OTT players, ie the relationship with the consumer. Consumers will gravitate towards simple, trusted providers and solutions so if mobile operators could offer competitive cloud-based services that were integrated in the devices and the billing, plus add the reliability of the network, they will come.

Carriers already dominate the communications (and to some extent the personal media distribution) value chains. Energy and government service provision are certainly important areas to study, but in most cases Business IT seems a good place to start due to the general trend of SMEs moving to opex oriented models for software, systems and services.

By leveraging their network assets, operators add value by exploiting user attributes such as profiles and activities, making cloud services relevant and meaningful to users and providing the linkage between the upstream and downstream components of multi-sided business models.

Mobile Cloud Computing enables these payments to be made through customers' phone accounts . Furthermore plug-ins can take advantage of NaaS features. One of the most valuable services, that can enfranchise many thousands of new ASPs, is the 'Bill on Behalf-of' (BoBo) ability for them to be paid through their customers' phone accounts. Comprehensive payments and settlement and the associated business infrastructure, is a critical component of Mobile Cloud Computing.

For telecom operators offering enterprise cloud services, target segments can be based on enterprise size for horizontal applications, such as unified communications (UC) and enterprise resource planning (ERP), or it can be vertical market-based for industry-specific applications such as meter data management for utilities companies. We can expect industry-specific cloud services to benefit from the rise of M2M communications, and vice-versa, and there are already cloud offerings purposely built for airlines, healthcare providers and financial institutions.

The requirements for implementing Mobile Cloud Computing are three pronged :

a. Developer friendly standardised interfaces

b. Appropriate carrier grade platforms and architecture, which support both the highly technical communication infrastructure and secure and auditable payments, settlement and charging infrastructure

c. Cross network service providers to establish the necessary business relationships with operators and with each other (for inter-region access). The GSMA's OneAPI commercial pilot in Canada is paving the way for this.

The global mobile industry's main trade body, the GSM Association (GSMA), has recognized the need for (mobile) telecom operators to treat their assets as marketable resources and to offer APIs to third parties, i.e. to embrace the Network As a Service model.

According to Gartner, the single biggest revenue opportunity for CSPs is as a Cloud Service Brokerage. This concept is beyond product bundles, it is creating an aggregation and integration of ICT solutions from a single source. The cloud application marketplace is loaded with solutions, creating confusion and challenges for the cloud service providers and consumers alike.

Becoming a cloud service brokerage, and bundling those services with their existing traditional telecom services has many advantages. One of the main benefits is service velocity, as XaaS services (Software-as-a-Service, Infrastructure-as-a-Service, etc.) can be on-boarded for pricing, fulfillment and billing in weeks, as opposed to the typical months-to-years timeframe. In addition, the costs to implement the services are dramatically reduced as well.

---♠--

4.7 : Telcos : get LEAN and MEAN not EXTINCT

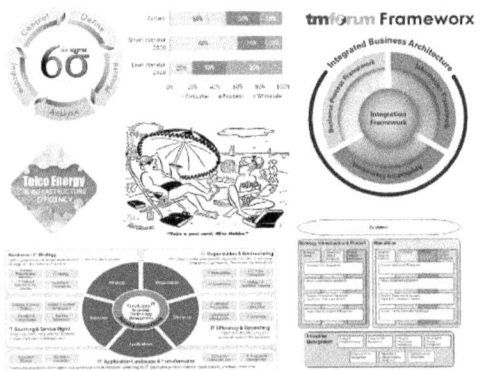

Telcos are under constant pressure to optimize operational costs, gain agility and offer superior services to customers. In a wicked competitive environment , containing costs, streamlining operations, retaining customer loyalty, and maximizing the Average Margin Per User (AMPU) becomes a business imperative. Short product life cycles and overheated marketing are overwhelming the operators, which resort to ad hoc solutions that appear to offer customers what they want, but in fact mask additional costs. These costs, however, may not show up until further down the service delivery chain in other areas of the business, where their root cause may be understood but cannot be addressed across functional boundaries.

Complexity is a fact of life for telecom operators, but it is also a cost driver. Legacy systems are maintained alongside next-generation networks .The complexity that has overtaken the telecom business has resulted in organizations with technology frameworks, tariff structures, and product catalogues that if plotted on a chart would resemble a Jackson Pollock painting.One European operator found that it was offering 20,000 different tariffs to 15 million customers in one country; after it streamlined its processes to respond to real customer needs, the number of tariffs was reduced to 8,000 !! Analysis by the BCG shows that Telecoms is one of the most inefficient industries with over 40 % of its cost base gobbled up by waste in various telecoms processes.

So what do we do ?? To start with look at TM Forums's Business Process Framework (eTOM) : a widely deployed and accepted model and framework for business processes in the Information, Communications, and Entertainment industries. As a key part of TM Forum's Frameworx, the Business Process Framework represents the whole of a Service Provider's enterprise environment in a hierarchy of process elements that capture process detail at various levels.The Business Process Framework (eTOM) describes and analyzes different levels of enterprise processes according to their significance and priority for the business . For CSP's , the Business Process Framework serves as the blueprint for process direction. Here are some case studies to vindicate e TOM's effectiveness.

Qwest wanted to transform its service delivery to shorten the time-to-market for new products, including cloud services, reduce its operating costs, and have visibility and traceability from products to services to resources. It was also determined to reduce individual service component redundancy and enforce Qwest's high standards for the overall customer experience.To reduce investment risk and prove the viability of what it wanted to achieve, the operator and its partners turned to TM Forum's Frameworx and Catalyst Program before it embarked on the transformation. Within a year of the deployment Qwest saw a 4 percent increase in revenue, a 5 percent cost reduction, a 25 percent improvement in new product deployment cycle times, and a decrease in unique provisioning and assurance job steps.

Magyar Telekom's project to convert a legacy provisioning system into a single platform successfully enables the provisioning and activation of multiple product lines. The implementation relied heavily on TM Forum Frameworx and is delivering many benefits. They include cutting service activation by 20 percent and increasing the ratio of successful automated activations by 30 percent. Time-to-market for services was reduced by up to 20 percent, while the time needed to integrate new network management systems fell by 30 percent. When manual interventions are needed, they take 70 percent less time. The deployment of a zero-touch home gateway has lessened field force activity by 30 percent. New and existing services are being migrated to a new platform, and CRM will be enhanced to support trouble ticketing and the management of service level agreements.

Concurrently with e TOM an approach called Lean Six Sigma (used so succefully in corporates from the Fortune 500 financial , manufacturing and service industries) can be implemented to cut waste and inefficiency in Telco processes. Lean Six Sigma is a managerial concept combining Lean and Six Sigma that results in the elimination of the eight kinds of wastes / muda (classified as Defects, Overproduction, Waiting, Non-Utilized Talent, Transportation, Inventory, Motion, Extra-Processing) and provision of goods and service at a rate of 3.4 defects per million opportunities (DPMO).

Lean Six Sigma utilises the DMAIC phases similar to that of Six Sigma. DMAIC (an abbreviation for Define, Measure, Analyze, Improve and Control) refers to a data-driven improvement cycle used for improving, optimizing and stabilizing business processes and designs.

There are 4 overarching strategies in the endeavour to create a LEAN MEAN Organisation.The categories can be seen as structural, transformational, changes with high complexity. Pursuing any of these should not be seen as a replacement to the first strategy of continuous improvement – there is always something more that can be done to improve the efficiency within the business as it is today.

Improve cost efficiency and productivity through automation, centralisation, market differentiation and reengineering of work processes (including partnering)
Realise national economy of scale by mergers & acquisitions with competing operators (including network sharing)
Achieve international economy of scale by implementation of cross-border working processes
Leverage national economy of scope by integrating fixed, broadband, TV or mobile businesses

Today, every aspect of your telco's operations needs to be measured against the touchstone of COST EFFICIENCY to ensure it brings profits. Whether it is investment in Transformative IP programs, in merging and acquiring enterprises, or in outpacing competition; eliminating redundancies and optimizing processes is essential. Getting rid of people (to cut costs) hardly requires imagination unless senior execs have overloaded the Company with relatives , buddies and assorted PA's !! Using both e TOM and Lean Six Sigma Telcos can cut costs INTELLIGENTLY in a variety of areas as identified by the gurus at AT Kearney :

Network, marketing, and IT : These three areas have the most potential for optimizing operational and capital expenditures, typically by reducing complexity.

Supply chain and procurement : Some Global Telcos aspire rapid international growth—often through acquisitions presents plenty of opportunities to improve supply chain and procurement capabilities. By standardizing purchasing requirements and internal technical specifications, consolidating volumes, and optimizing deals with suppliers, operators can cut costs without affecting core operations.

Back office : Consolidating back-office functions such as HR and finance, potentially by establishing central or regional shared services, can increase efficiency

Information technology : Centralizing IT services and standardizing or consolidating applications and hardware can substantially reduce costs and often improve service.

Infrastructure sharing : Sharing infrastructure among operators is another way to optimize costs and leverage economies of scale. For example, Bharti, Millicom, and Vodafone (Spain, Germany, U.K., India, and Ireland) have shared networks with other operators. In Sweden, 3 and Telenor's joint venture, 3GIS, covers around 70 percent of its network with shared infrastructure.

Outsourcing: Outsourcing non-core activities, such as fleet services and facility management, can improve efficiency and allow more management focus on customers. Newer outsourcing models include managed capacity, where an outsourcer is paid on a variable utilization or capacity basis. These models, besides increasing efficiency, reduce risk, and limit financing needs while fundamentally shifting the focus from operations to customer experience and partnership management.

Energy efficiency : Energy efficiency can cut costs while reducing environmental impact. France Telecom-Orange, for example, is aiming to reduce energy consumption by 15 percent between 2006 and 2020. By the end of 2010, the group had fitted more than 8,000 network sites with optimized ventilation systems, cut energy consumption at data centers, and installed solar-powered base stations (mainly in Africa and the Middle East).

Telenor, for example, reduced its software licensing costs by 34 percent by replacing local licensing agreements with global deals.Telcos will need to use the full scale of their groups to create synergies, reduce external spending, and benefit from solid supplier relationships, which can bring earlier access to new handsets and network equipment.

Bharti Airtel's so-called "Minutes Factory" has enabled it to target millions of pre-paid customers that would have been too costly to serve using the conventional subscriber-led model. The factory's key elements include outsourced network equipment, which enables fixed costs to convert to variable costs. Bharti's partnerships enable it to add network and IT capacity quickly and efficiently, as needed.

We know that taming complexity and streamlining operations can reduce operational costs by a third and provide customers with better service. The up-front savings achievable in the short term—six to 12 months—will cover the costs of the initial assessment that identifies how and where to implement a Lean transformation using eTOM and Six Sigma methodologies.The good news for telecom companies faced with stalled revenue growth is that there are ways to significantly reduce expenditures.

Telecom carriers will have to lower their operating expenses for traditional telecom services to maximize free cash flow, which can be invested in nontraditional services. Telcos must focus on operating efficiency when offering a suite of non traditional services in the 4G data world services, as there are no "killer" applications.

Operators that do not undergo a lean transformation, however, will find themselves unable to compete. The decision to adopt a lean and mean approach needs to be made before you become extinct !!

---♠---

Chapter 5 : Craft the winning business model

5.1 : White Space : Digital Ecosystems in the New Mobile Web

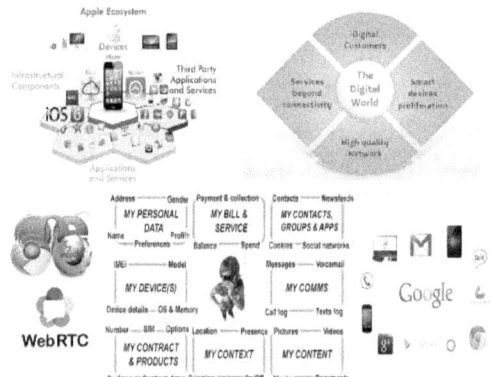

The Telco 2.0 Initiative Gurus believe that the emergence of the New Mobile Web is creating a white space for operators seeking to (re)build their role in the digital marketplace. The New Mobile Web is a term used to describe the transformed mobile Web experience achieved through advances in technology; HTLM5, faster, cheaper (4G) connectivity, better mobile devices.

The New Mobile Web will lead to a shift away from native (Apple & Android) app ecosystems to browser-based consumption of media and services. Web RTC is a technology that promises to bring voice and video into the web browser. At the same time, it offers an excellent opportunity for Telcos to innovate by extending telecom services into the open web and freeing voice from closed telecom networks.

"Software is eating the world," says Marc Andreessen, co-founder of Netscape, who himself was once on the receiving end of competition from Microsoft. For most industries it's not a matter of 'if', but a matter of 'when' you will face a software-driven competitor with a unique business model.Low (or zero) marginal cost of many modern technology products is the key enabler for the success of new age business models. This is why these business models (pioneered by the software goliaths)are different from traditional loss leader and bundling tactics.
As these business models are self-sustained, the new age competitors fundamentally alter the market dynamics in the "victimized" markets. The new market conditions are often rendering the business models of incumbents unsustainable.

Lured by the promise of attracting higher-ARPU smartphone users, Telcos have worked hard to flood the market with smartphones at a wide range of price points. This strategy served the short-term goal of boosting the connectivity business, but may have jeopardized the long-term competitiveness of the service business by surrendering the customer ownership associated with authentication, user identity management and billing services.

As the basis of competition changes to "choice and flexibility", vertical integration (ie : notion of all-in-one telco spanning network operations, telephony, messaging, data access, user identity management, authentication and billing, as well as distribution and retail) has lost its competitive lustre. The lack of flexibility inherent to vertical integration explains why Telcos lost out to smartphone and Internet platforms in the areas of location services, authentication, single sign on, user identity, and billing.

Although many Telcos believe that they urgently need to build strong digital businesses, most are struggling to do so. Creating a

Digital Telco means looking beyond traditional telco business models in the context of the changing telecom value network.The challenge for Telcos isn't that OTT companies outspend or " out imagine " them in digital innovation. It's that marginal cost analysis steers Telcos towards investments in capabilities that were relevant in the old basis of competition, rather than toward developing new capabilities relevant for the new basis of competition.

For example marginal cost analysis makes RCS (e ,5) backboned on IMS is an attractive choice for new presence and messaging services designed according to traditional telco service models. However, according to the new basis for competition, the scalability and interoperability offered by IMS are less important than flexibility. Telcos could be better off investing in new, more flexible infrastructure better suited for experimentation with new services, use cases and business models. As such the focus of Web RTC innovation should be on building developer ecosystems for voice services, discovery of new use cases and experiments with new business models, and not on technology. If Telcos won't do this, competitors will.

Imagine for a moment that you run a mobile operator who derives 10% of its revenues from SMS (which commanded 70% margins in the heydays of mobiletelecom). In a matter of few short years, these pesky free messaging apps destroyed your most profitable business. At first you thought that these "free apps" would go away after they burned through the money provided by the venture capitalists. Then you tried to compete with homemade competitors like WAC and Joyn, only to discover that it's a sure recipe to lose money.

Unfortunately, since the new generation of messaging apps have unique business models, they can sustain free services indefinitely and forever change the dynamics of the mobile messaging market. WeChat, Viber, Line and others monetise by using their platform for e-commerce, selling digital goods (stickers, games), physical goods and services (like taxi rides). They don't have to charge for the messages or even voice calls to be commercially successful.

Opportunities in digital content fall into two broad categories: production and distribution. In IBM's opinion, content production offers little potential for telecom providers; most operators will do better by partnering with content providers than by attempting to produce content themselves. However, they can also play a role in facilitating the trend toward user generated content by enabling consumers to enhance their own content with a range of telecom capabilities, including location, presence and interactive services.

To defend and grow their share of the digital content market, Telcos will ultimately have to make a substantial organizational, cultural, technological, operational and business model transformation as they transition from providing network connectivity to enabling the consumer's digital experience. With the transition to Internet Protocol (IP) and the proliferation of content distribution platforms, consumers increasingly want choice, flexibility and control over the multi media multi screen experience. Telecom operators can draw on their unique skills and capabilities to capitalize on this trend and distinguish themselves from rival platform providers.

Vision Mobile states that Telcos must move their innovation focus from technologies (be it HTML5, NFC, IMS, VoLTE, M2M or RCS-e) to ecosystems. Ecosystems are much better at delivering choice and flexibility.Ecosystem economics are driven by network effects and lock-in. iPhone apps attract Apple users, who in turn attract more developers, who make more apps, which attract even more users, and so on. This network effect between developers and users drives the explosive growth of the iOS platform.

Lock-in creates natural "walled gardens," as users develop habits around apps, while developers are locked-in by high switching costs created by their investments into the platform. A platform business model is about leveraging an operator's underutilised, walled network assets, taking a cut from the delivery of innovative services, in the same way that Apple takes a cut from the delivery of mobile apps or Facebook takes a cut out of ad delivery.

Over the longer term, Telcos can look for ways to build parallel ecosystems, using lessons from the ecosystem economics textbook. M2M holds the potential to create a vibrant ecosystem of users and solution providers, thereby establishing strong network effects and lock-in. Telcos can become the central force in this emerging ecosystem if they learn to engineer the ecosystems to their advantage. By looking at M2M through the lens of ecosystem economics, operators will see opportunities that are much bigger than just selling modems and data connections.But when it comes to making strategic decisions, the digital leaders have proventhat there is no reason to be bound by this artificial framing of markets.

The study of markets worked well for industrial age but in the Digital Era companies can get unfair but deserved advantage by smashing industry boundaries : That is by competing across multiple markets at the same time.Apple created an unfair advantage by competing in both consumer electronics and digital content markets. Google created an unfair advantage by competing in both online advertising and mobile markets. Amazon created an unfair advantage by competing in both e-commerce and tablet markets. Their direct competitors (Nokia, Yahoo, eBay) didn't stand a chance as the mobile revolution unfolded.

In the future, our communication will revolve around social media platforms, dominated by Internet firms such as Apple, Facebook and others that do not even exist yet. The Google+ video telephony project launched in 2011 shows the way forward for integrated communication environments that, building on a personal or organizational network, set up telephony, messaging, mail, chat or video links at the click of a button : Hey .....Customers don't buy access, they buy service ecosystems !!

--♠--

5.2 : Scenario Planning : Alternative futures for Telcos

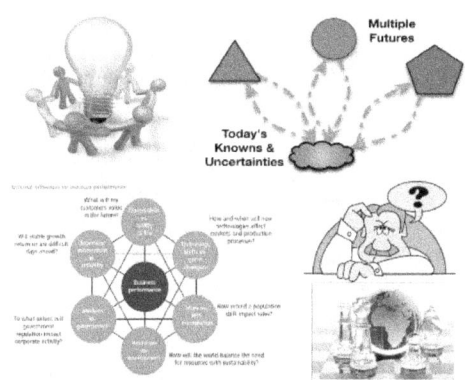

Scenario planning, is a strategic planning method to devise flexible long-term plans. It is in large part an adaptation of classic methods used by military intelligence : WAR GAMES. At Royal Dutch/Shell it was viewed as changing mindsets about the exogenous part of the world, prior to formulating specific strategies. Scenarios provide a mature insightful way to lay out alternative futures to challenge cherished but perhaps irrelevant assumptions on the future of the high capex volatile industries like Oil and Telcos.

The aim of scenario planning is not to identify one specific future, but to explore possible futures and to identify the mechanisms that drive the future in a particular direction. Scenario-based strategic planning achieves its cognitive effects through a structured, tool-based approach .Which brings us to envisioning some simple yet plausible scenarios for the Telco Industry !!

Scenario 1: The technology future

In this scenario Telcos obtain a competitive advantage by developing / delivering new proprietary technologies. At the same time customers' buying decisions are mainly influenced by the technological characteristics of the innovations. And consumers have become technology savvy and technology addicts. Under these circumstances, the most relevant strategy for telcos is to invent new technologies and push them to the market.

In the long run this scenario may mean the disappearance of fixed networks as the leading technology for transporting voice and data. Even if physical networks remain, in this scenario they will play a minor role. They are what roads are to transport: they need to be there, they need to be maintained, but apart from that nobody really cares. The customer is interested more in cars (Smartphones), than in the type of asphalt (LTE 4G) he drives on.

For example Google is highly innovative, only distributes services online and has completely disrupted existing business models, putting pressure on such diverse industries as traditional software, newspapers and publishers. In the technology world, telecom companies that are disruptive innovators will do to the current telecom industry what Google has done to software. They will redefine existing industry boundaries completely, based on technology.

Scenario 2: The commodity future

In this scenario there are no distinctive technologies that give telcos a competitive edge. Instead, there are a number of standard technologies offered by vendors. Consumers recognize that technological change is incremental : they are all very much aware of the standards that exist and have a pretty good idea about what they want. They do not need to go to a store to check the hardware, but just order it online. Telcos will have to offer the technologies and the connected services in a convenient and efficient way.

Gaining economies of scale to serve mass markets cheaply is the key to company survival here. The role of the Telco is best described as a standard direct packager: it takes standard technology and services from other companies, then packages them and offers them on the web with little thinking about product differentiation.

For example Dell does not innovate its technology, but buys components from vendors. Next it ties those components together and ships the resulting product fast and cheaply to the consumer. It is a completely online business model. As most people now have an understanding of what they want from a pc or laptop, or they have a smart cousin who can help them articulate their wishes, there is no need for a store with staff explaining what the basic choices are.

Scenario 3: The customised future

In this world each segment has its own demands in terms of products and services and how these are to be delivered. R&D focuses on serving this diversity of market segments. Telcos will follow a demand pull strategy in developing innovations. They will listen to the market, identify the needs of different consumer groups and innovate around this need. This may lead to different versions of products and services being offered in the market or to entirely different technologies being developed for different segments.

The role of the Telco is that of a market driven innovator which may lead in two directions. The first one is reactive: creating innovations based on a profound understanding of consumer needs. The other road is more proactive: creating innovations that create new market segments that other telecom companies have not yet thought of.

For example Apple innovates based on superior understanding of what consumers like. In doing so they deliver superior technology to existing markets (the iPod that replaced other MP3 players), but they also create new market segments by supporting online communities. Their approach is decidedly multichannel with much online business, but also investments in Apple stores and resellers.

Scenario 4: The segmented future

In the segmented world, telcos source innovation from vendors and package those standard technologies for specific market segments. In many ways this scenario is closest to the current strategies of most traditional telecom incumbents. They have decreased their investment in innovation and have embarked on a multichannel strategy. In the segmented world technology and content sourced from vendors need to be integrated and they need to be matched with the diverse market segments that exist.

The company that is best able to connect the trinity of technology, content and segment will be able to reap superior profits. However, a critique on this strategy is that it tries to be everything for everyone. There is little focus on either specific technologies or segments. The risk that niche players enter into the most profitable segments may be high. Have Telcos ever considered the nightmare scenario when Apple and Amazon offer handsets plus e-SIMs. Contracts and billing could be handled via iTunes or Amazon accounts. The telco would thus become anonymous, a mere wholesaler of network capacity with no end-customer relationship of its own.

Traditional strategy tools are designed for stable market environments, but fail predictably when applied to innovation under conditions of uncertainty and rapid change, which characterizes today's telecom market. Telcos can no longer afford the luxury of sluggishness, strategic inertia or blind spots.The good news is that various scenarios show an abundance of opportunities for Telcos.

The mobile industry is undergoing a dramatic rethinking of business foundations and supporting technologies. In many ways, technologies such as cloud, software-defined networking and 5G result in a "software is eating the network" end game. This in turn will promote opportunities that are much larger than just selling voice and data access meaning digital commerce, advertising, energy services, smart home , e-health ,M2M , Connected Car , Big Data , IoT etc .It is for the Telcos to adapt , improvise , transform to profit from these opportunities ....lest Telco CxO's forget what the 6 th century Sage said......... " Plants are born tender and pliant; dead, they are brittle and dry ....and the same for men " Lao Tzu

--♠--

5.3: Strategic Insight : Blue Ocean path for Telcos

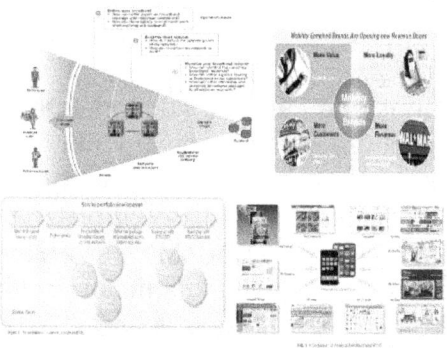

Essentially, the future of the convergent industry is in service provision with globalization and personalization spurring consumer demand. The upsurge of convergence has brought not only opportunities, but unparalleled resource advantages to operators. However, whether an operator

can go with the flow and thrive in the new environment depends not only on reformulating strategy but also on its timely and effective implementation. Ever since strategic management was introduced to economy field, its main purpose is pursuing competitive advantage in the existence market. Blue Ocean strategy is a totally new strategy compare with old ones.In their landmark book strategy gurus Kim and Mauborgne (2005) divide the market universe into two parts: red oceans and blue oceans.

"Red oceans " is described as all the industries in existence today which is the known market space. " Blue oceans " refers to all the industries not in existence today which is the unknown market space. In red oceans, the market boundaries is clearly identified, the market space gets crowded, prospect for profits and growth are limited, competition in red oceans turns to be bloody.

 In contrast, blue oceans are defined as a new space in which market boundaries and industry structure are not given and can be reconstructed. In blue oceans, competition is irrelevant since the rules of the game haven 't been set. Very few incumbent telecom providers has put into place any Blue Ocean Strategies. Yet Blue Ocean Strategies have made the Circus, Wine, Gaming, Airline, etc. industries exciting again, so why not apply it to the telecom market ?

Blue ocean strategy is about creating and capturing uncontested market space, thereby making the competition irrelevant. For example, NTT DoCoMo was the first company to make money out of the mobile internet. In a very competitive industry engaged in a technology race and strong price erosion, NTT DoCoMo was able to achieve superior performance when it launched its novel i-mode services in February 1999. It was an immediate and explosive success in Japan.

As with NTT DoCoMo, the goal for a firm's blue ocean strategic move is the pursuit of value innovation — a leap in value for buyers and company alike. This comes from simultaneous pursuit of differentiation and low cost. NTT DoCoMo pursued the concept of Value innovation : a new way of thinking about and executing strategy that results in the creation of a blue ocean and a break from the competition. More importantly, value innovation defies one of the most commonly accepted dogmas of competition-based strategy — the value-cost tradeoff.

In order to achieve its blue ocean strategy, KPN (the largest Operator in the Netherlands) has mapped out four concrete objectives: fixed and mobile service convergence; full utilization of current network resources; considerable decline in CAPEX and OPEX; and delivering a variety of new services, which are built on an All-IP network. These services would include: multimedia personal communication services such as, voice, video, photo, data, message, and PTT; IP corporate communication such as, an IP private line, IP PBX, and multimedia service; and multimedia entertainment such as, games, IPTV, the worldwide Web, and Portal.

Operators need to realize that extending connectivity alone cannot keep them afloat. Instead they require software, device and service strategies that can add value and at the same time differentiate them from competition.In the future, the primary operational mode for large players might well be as aggregators of massive services. The key is to open the platform and gain as much partner power as possible. This is the fundamental reason why concepts like Web2.0 and P4P become important.

So what is a blue ocean opportunity in mobile broadband ? Could it be a PPP ?Based on smartphones a PPP (Personal Portable Portal) is generated by integrating frequently used personal applications into a terminal platform for easy and quick access. These applications include vertical portals, professional services, communities, directories, personal information management (PIM), location-based services (LBS), work, entertainment, activities, images, IM, sports, music, reading materials, and Emails.A PPP is distinguished by its inherent freedom, flexibility, and adaptability. Its features include a framework, user data hosting and management, virtual resources, and independence from terminals. Personal portals are easily configured, combined, updated on the web, and loaded onto various terminals.

Many companies, particularly technology firms, do tend to continuously add small features to their products in an attempt to differentiate themselves from the competition through a continual process of "incremental innovation." Mobile telephone companies are particularly guilty of this, yet each additional "design feature" detracts value from the buyer as the phone becomes increasingly difficult to use.

A study from Eindhoven University found that in the US nearly half of products returned by customers for refunds were in perfect working order, their owners just couldn't figure out how to use them. Innovation without value tends to be technology-driven, market pioneering or futuristic design that may shoot beyond what buyers are ready to accept and pay for. Value without innovation tends to focus on value creation on an incremental scale that, at best, improves value but is not sufficient to stand out in the market.

For Telcos , most value customers are real-name customers, unlike Google's anonymous and even nameless customers. Only operators can obtain data on user behavioral patterns, which is a huge advantage over other competitors.Hence, operators are in a position to help their partners promote their products and recommend suitable products to the right customers. They can match massive services to millions of users. This is the service aggregators core competitiveness, yet most operators have been asleep at the wheel and have not collected, collated or utilized the data.

Many applications, such as home security protection, health and medical care, do not merely involve information distribution and interaction, but require hardware deployment and maintenance as well. The OTT providers find it hard to provide delivery or services in a large area and operators with massive service teams can help them and profit as well. The OTT providers lack branding power and a strong credit rating. When users purchase their services, a guarantee is needed.

Operators with long-term operations experience and a good credit and credibility rating can play the role of guarantor. This means that operators must qualify suppliers and control risk. Similarly, when users pay for a service, they usually will not trust an unknown service provider. A third-party, reputable platform is needed for completing settlement and payment actions. This is where operators' existing mature billing platform can really shine.

Salesforce.com's strategic moves provide an exemplary demonstration of how a company can effectively create and renew its blue ocean in the B2B field by value innovating its single business on the product, service and delivery platforms alternately. By 'de-segmenting' the market and looking at exceptional buyer value across segments and looking for commonalities across non-customers, Salesforce created a new mass of buyers that traversed the traditional segment boundaries. Had they wished to make incremental changes in the industry they would probably have offered a traditional client server model with some element of web-based access as well.

Salesforce decided to launch a new concept around the mantras of 'success not software' and 'low cost, good enough', which completely reinvented the way the industry thought about CRM and challenged the traditional value-cost trade-off that buyers were typically used to. In keeping the first principle of Blue Ocean Strategy, Salesforce broke out of the conventional wisdom trap and pioneered to create a new value proposition that forced the market to wake up and listen.

Pureplay broadband access – the mainstay of traditional telcos' business today – will remain the cornerstone of digital communication in the future for both landlines and mobile communication.But Telcos should build an open digitised platform, utilize their long-term operational resources and experience to integrate more and better services and provide desirable services to end users.

By doing so, operators can gradually move their commanding positions from network connections to their own exclusive services or the services with the best user experience.The "right" strategic orientation for each telco depends on five key levers:

1) Personalization of service ecosystem and the customer experience
2) Uncompromising defense of relationships with end customers
3) Cost-efficient broadband network build-ups
4) Realignment and radical streamlining of operating models
5) Financial resources to drive digital transformation and consolidation.

If telcos do their homework with respect to broadband access strategy, customer experience , transform themselves into a cost efficient lean operating model backboned on Omni-digital,they have a real fighting chance to recover the ground lost to the OTT players ...instead of complaining they have been relegated to " dumb pipes ".YES ..you are dumb if you don't learn how to swim in the Blue Ocean.Whatever the future holds, all Telcos are called on to refocus and realign – and to do so fast !!

--♠--

5.4 : SingTel : Red Dragon of the Telco World

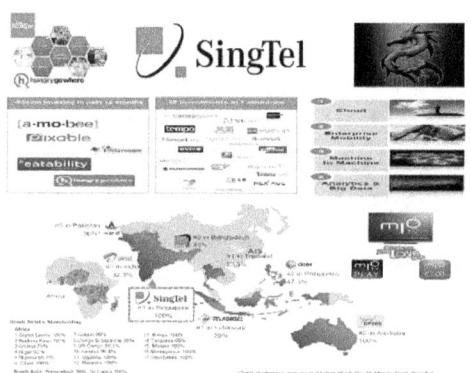

There is a small island nation without any natural resources where its 5 million inhabitants generate an explosive economy worth $ 300 + billion with a GDP per capita of $ 60000 +. They are considered one of the most wealthiest nations in the world and a testament to a culture of enlightened discipline and personal rectitude . Yes : it is Singapore !! And this powerhouse nation is also home to a Telco Titan : SingTel

SingTel is Asia's leading communications group with operations and investments around the world .With significant operations in Singapore and Australia (through wholly-owned subsidiary SingTel Optus), the Group provides a comprehensive portfolio of services that includes voice and data solutions over fixed, wireless and Internet platforms, as well as infocomm technology and pay TV. The Group has presence in Asia and Africa with 473 million mobile customers in 26 countries, including Bangladesh, India, Indonesia, the Philippines and Thailand.

SingTel's domestic operation, is a typical full-service converged carrier. Currently, their top priority is the deployment of 4G wireless and FTTH in the fixed sphere. SingTel is cooperating in the Singaporean government's national broadband network project, which aims to provide 1Gbps service and to offer full openness at Layer 2 and also at Layer 0 (i.e. to the ducts).SingTel are also a very significant force in carrier services, wholesale, and IP transit. SingTel is also aiming to double the size of its satellite business, with two satellite launches scheduled within the next two years via JV's.

Recently the operator decided to restructure its business into three units : group consumer, group digital life and group ICT-in order to sustain growth, competitiveness, and innovation. With the reorganization, SingTel plan to reinvent its core carriage business, create and drive new growth platforms that leverage and strengthen the core, and turbo-charge its regional capabilities in ICT services. They broke new ground with the introduction of PowerON Compute Service. This state-of-the-art cloud solution provides enterprises with the business agility and cost effectiveness of public clouds without compromising on portability, compatibility, security and control demanded by enterprise IT organisations.

SingTel is banking on acquisitions of smaller companies to help drive growth in its business as revenue slows from mature markets like Singapore and Australia. SingTel plans to spend US$1.6 billion in three years to acquire companies specializing in digital advertising, content and entertainment. That $1.6 billion investment will be spread over the next three years, and will be largely ploughed into strategic acquisitions in the online space that can tie in with SingTel's phone services across the region. SingTel spent US$400 million in acquiring advertising, entertainment and digital commerce firms in the last fiscal year, including the US$321 billion it paid for Amobee, a U.S.-based mobile advertising company. By leveraging their unique assets and Amobee, they will be able to realise the full potential of mobile marketing as a platform to change the way brands communicate with their customers.

The company is also using surplus cash to step up dividend payments to keep investors happy even as it continues to invest in its core telecommunications business. SingTel's dividend payout ratio ranges from 55 per cent to 70 per cent of underlying net profit. The Group will continue to review at least on a three-year basis its cash needs for operations and growth, with a view to returning surplus cash to shareholders. This is consistent with the Group's commitment to an optimal capital structure and investment grade credit ratings, while maintaining financial flexibility.

Companies like SingTel go beyond access business, positioning themselves as service providers and complementing their traditional sales and network operations with a third element, a "telco innovation factory" charged with developing and marketing new services. The innovation factory consist of access-centric services that use the existing network and IT platforms – in the e-health segment, for instance – or regional OTT-related offerings such as TV. Such services will be embedded in partners' service suites or, depending on the extent to which the ad valorem aspect is to be emphasized, on proprietary platforms that integrate third-party services.

SingTel is attempting to buck the trend of telcos becoming just a big, fat dumb pipe that only competes on price. Their vision is to really go after the heart and soul of the consumer, ultimately to drive a deeper connection that substantially increases the value of SingTel.

Aligning strategic positioning with shared values opens up fundamentally new ways of thinking about business. From being known as a telecom carrier, Singtel can become a smarter cities builder, a wellness provider or a health and safety enabler : whatever !! They have recognized that mobile operators will play a crucial role in working together with a range of industry partners in health, automotive, education, smart cities and a range of vertical industries to accelerate the launch of valuable connected services.

SingTel Life Labs is a global innovation initiative to accelerate innovation and application development through collaboration with strategic partners, renowned research institutes as well as the innovator and developer ecosystems. They created an App that helps Singapore residents navigate increasingly large and confusing shopping malls using " sensor fusion " and Wifi triangulation .

Chua Sock Koong, SingTel Group CEO, states : "SingTel has a long history of quietly, but successfully, making bold and industry-shaping investments. We now see some of the largest and most exciting opportunities that have ever existed in this industry... We need to learn from past failures and be prepared to reiterate a bold idea if we believe it will eventually bear fruit. Even when we do not succeed, we expect a "fail fast and fail cheap" mentality to produce valuable learnings that can form the basis of long-term advantage against competition.."

Strategic partnerships , savvy aquisitions , superior investments , superfast super rich broadband networks : this kind of cerebral leadership summarises why SingTel has become such great company. And by the way ...Chua Sock Koong is a female CEO in a male dominated telco world !!!

---♠--

5.5 : Telco M/A Reloaded : the anatomy of a good deal

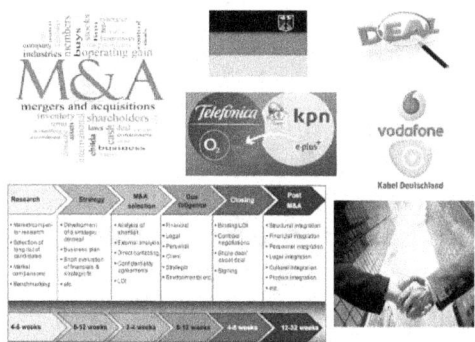

Mergers and acquisitions have been a mainstay of the telecom industry for many years. In the past decade, the telecom industry has spent an astounding USD 1.5 trillion + on M&A activities, investments that have transformed the industry landscape into the competitive playing field we see today.

McKinsey reckon that over 80 percent of aggregate deal value has focused on core telecom services and spectrum license market segments, with the remainder targeted toward adjacent markets such as infrastructure, connectivity providers, multimedia and financial services.The twin requirements of nonstop technological advances and ever more capital to pay for them produce an inherently dynamic marketplace that will keep the telecoms industry at the forefront of M&A action around the globe.

Wealthy telcos from Middle East and China are on the prowl for well-priced targets both within and outside of their home markets, and in general strategic operators will outbid private equity players for attractive targets. The continued tension between growth in allied fields (for example, telecom operators acquiring cable TV companies or application providers) versus specialisation in one subsector (telecom operators spinning off cell towers) will also serve as kindling for M&A activity throughout the sector.

Vodafone agreed to pay 7.7 billion euros for Kabel Deutschland, the country's largest cable company, because it combines phone, Web and TV services to increase customer loyalty and stabilize prices.

In one of the largest M&A deals this year was Telfonica's agreed to buy the E-Plus German wireless unit of Royal KPN NV (KPN) in a cash-and-stock deal valuing the unit at 8.1 billion euros ($10.7 billion) to become the country's biggest mobile-phone operator by customers. The Dutch phone company will get 5 billion euros in cash and a 17.6 percent stake in the combination of E-Plus and Telefonica Deutschland Holding AG (O2D), the Spanish carrier's German unit, which uses the O2 brand. As the EU commissioner in charge of the digital agenda, pushes reform in favor of a single European telecommunications market, carriers have become more emboldened to pursue deals. They are seeking to share expenses to build so-called fourth-generation networks to cope with rising demand for faster data connections

An analysis of the transactional rationale of the Telefonica deal provides valuable insights into the main elements for considerations of " best of breed " Telco M&A deal meaning covering the financial and non financial bases. Ofcourse there is no substitute for old-fashioned focus on the fundamentals of M&A: a clearly articulated and well thought-out strategic rationale for the acquisition becomes the yardstick by which to measure individual decisions that arise during the course of a transaction. Without one, decisions are made that end up being costly and inconsistent with the ultimate strategy chosen – or worse, require divestment of the entire acquisition years later as a 'bad deal'.

First Telefonica's desire to create a Digital Telco Titan : become a leading player with a combined customer base of 43m, 42% in postpaid and derive strong scale benefits with combined mobile revenue market share of 32% . O2 and E-Plus's combined customer base at the end of March would leapfrog Vodafone's

32.4 million and Deutsche Telekom's 37 million, according to data compiled by Bloomberg Industries. Germany has become the hottest battleground for telecommunications assets in Europe as demand for video and music delivered wirelessly and over the Internet increases, while voice revenue declines.

Second motivation was value crystallization through significant synergies. Telefonica is targeting NPV of synergies of €5.0–5.5bn, net of integration costs with projected Net savings from year 2 and Annual run-rate synergies of approx. €800 m; 75% of run-rate synergies by year 4. The deal will result in cost savings and revenue "synergies" of around 5 billion euros.

Telfonica identified achievable synergies by rationalisation of distribution network ; increased efficiency in customer service costs leveraging best practices and scale ; better channel management and reduced overheads ; focused rollout on one common nationwide LTE network and improved quality from 3G network consolidation ; backbone, backhaul and core network consolidation, with reduced OpEx from network integration (rentals, power, maintenance, transport costs, overheads) ; site consolidation and rationalisation via reduction of around 14,000 sites ; increased efficiency by leveraging scalable transmission agreement with Deutsche Telekom and reduced SG&A expenses by process rationalisation and a focus on become a more lean agile organisation.

Third motivation Telefonica wants a single LTE network to provide what they call the Best Mobile broadband experience. Key factors included giving customers to benefit from the best high speed mobile and fixed experience from a single LTE network and access to future-proof DT NGA network ; Tariff innovation, voice & mobile data bundling ; Strong multi-brand portfolio across segments ; Offering ICT / cloud solutions for business customers ; extensive distribution channel and outstanding customer service ;Leverage convergence through cross-selling / up-selling opportunities as well as profiting from digital innovation and scale from Telefónica's global capabilities (data centres , portfolio of OTT services and partnerships) .

Final motivation was value Creation for Telefónica Deutschland Shareholders . Here they were looking for enhanced financial flexibility (improving leverage) while maintaining an attractive shareholder remuneration ; maintaining conservative pro forma balance sheet with a projected EPS and FCF accretive from first year of full operation. In addition the M&A was all about investing in future growth while reinforcing geographical diversification, increasing exposure to an attractive market with a positive impact on Telefónica's cash flow generation profile.

Telfonica opted for the " Financing Without Increasing Leverage " motto meaning the deal is very positive for Telefonica from a business perspective while it doesn't affect its debt position. They have a Rights Issue in enlarged Telefónica Deutschland of €3.70bn. Telefónica subscribes prorate to its stake of 76.8%, €2.84bn + €1.30bn to KPN for 7.3% stake in the enlarged Telefónica Deutschland. Required total financing of €4.14bn is structured as 50-65% Hybrid, 100% equity under IFRS/ 50% equity for credit rating agencies and 20-30% Mandatory Convertible.

Their objective is weighting around 2x incremental OIBDA, excluding synergies ; with Net debt/ratio preserved in short term for neutral to positive impact but keeping strong liquidity to maintain 24 months maturities for FCF generation till deal completion. Economic KPIs and cash flows must be consistent with real value creation. There is no place for speculation, particularly in these variable markets where sources of capital are skeptical, margins becoming tighter, and the consequences of missing forecasts are more direct.

As Ey rightly point out that a fanatical focus on due diligence of all aspects of the target's business and complete regulatory and market landscape is indispensable when there is so much money at stake . There is no substitute for old-fashioned focus on the fundamentals of M/A: a clearly articulated and well thought-out strategic rationale for the acquisition becomes the yardstick by which to measure individual decisions that arise during the course of a transaction. Without one, decisions are made that end up being costly and inconsistent with the ultimate strategy chosen – or worse, require divestment of the entire acquisition years later as a 'bad deal '.

Proactive Telcos clarify their M/A approach, organization and the way they source deals globally, and work closely with investor relations to ensure they have the right story to tell the capital markets—especially if they aggressively pursue new adjacent areas that have different value creation profiles or emerging market economies with majority low ARPU subscribers .

--♠--

5.6: Fast Track : Evolved Managed Services Reloaded

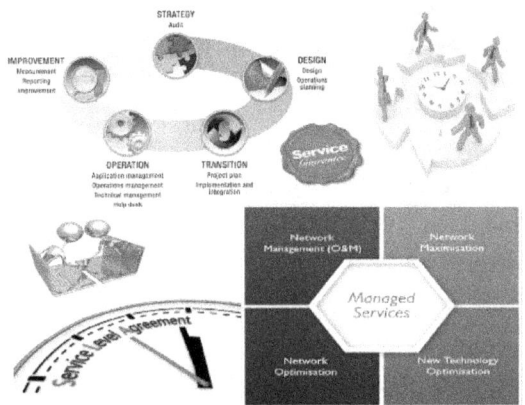 For many operators in
MEA Region, the cost of delivering services will increase faster
than revenues are keeping pace. In fact, this so-called revenue-
cost gap is generally a global phenomenon, as operators find that
single digit percentage reductions in expenses are simply not
enough when it comes to meeting customer expectations for
personalized multimedia services. As such the entire concept of
"managed services" has changed.

The complexities of running both fixed and mobile networks and
value added services continues to increase, while operators are
increasingly required to focus their scarce resources given a
highly competitive market. To maintain margin growth operations
have focused on cost-cutting exercises which has accelerated
outsourcing.

Operator's acceptance of partnering and acquisition of external
expertise has grown in recognition of the opportunities and
efficiencies such activities provide. Managed services has
emerged as an important delivery model for network equipment
vendors, systems integrators, BOSS specialists, applications and
service specialist as well as network-owning operators
themselves.

The main components of the old approach — NOC services, multivendor maintenance and network outsourcing — were sufficient in an environment where point solutions, cost savings and the bottom line were primary considerations. According to Analysts many companies are undercutting their own outsourcing initiatives by not managing them strategically. Most outsourcing initiatives do not attain the full desired benefits, especially cost reductions simply because Telcos fail to :

• Define a clear strategy of what activities should be outsourced, what objectives should be accomplished via outsourcing, and what boundaries should exist between any internal and external work. Aligning this strategy with the business units to maximize the value of any sourcing decisions is critical
• Establish a clear baseline of current costs (both the total and its components) and compare that to external benchmarks. Deciding what work can be optimized internally prior to outsourcing and use that as the basis to define cost targets with vendors is paramount
• Develop to establish a systematic demand management process that prioritizes project requests based on the proposed value creation for a business. And introduce consumption-based chargeback mechanisms to the business units to create an awareness of costs and to foster behaviors that are more conscientious toward usage.

Buyer organizations should use a structured, strategic approach to calculate the true costs of outsourcing, taking into account the risks and opportunities. If cost is a concern, then a clear knowledge of performance baselines before entering an outsourcing engagement is critical. To contain costs, companies should rely on benchmarks and adopt a consumption-based system that charges expenses back to the business units that incurred them.

Providers and buyers must work together to apply sourcing discipline and craft outsourcing relationships that address near-term cost objectives and longer-term scalability and enhancement. Clearly companies need to turn transactional vendor relationships into strategic partnerships by doing the following:

• Clearly define the objectives and expected results (KPIs) from any outsourcing partnership.
• Assign clear roles and responsibilities to in-house personnel for handling various aspects of vendor partnerships, from overall relationship management to individual projects and their specific deliverables.
• Agree on an interaction model with the vendor to ensure day-to-day communications occur, expectations are regularly aligned, potential problems are identified and addressed early on, and future opportunities, especially those involving new technologies, are identified.
• Flexibility is the operative word for outsourcing strategies and contracts because outsourcing can never be separated from business goals. To drive the desired results, outsourcing relationships must constantly evolve.

In the traditional BOM approach, a qualified and trusted partner can share a portion of the risk in the early years of deployment with the network operator. Instead of paying all CAPEX up front before realizing any new revenues, the network operator can share the risk of build-out with the partner by deferring a share of the cost until it realizes subscriber growth.

Since next-generation managed services are geared toward achieving business transformation via services-centric approach risk mitigation is evolving into a risk sharing approach. Risk avoidance will be an inhibitor for some companies to outsource, even those that could gain efficiencies from managed services and performance-based contracts with an external provider. Risk management best practice include a PLAN :

• To identify and evaluate technical and commercial risks due to outsourcing specific services.
• To take mitigating precautions wherever warranted, such as identifying alternate vendors for crucial pieces of work.
• To define and continuously monitor risk-related KPIs that trigger early action whenever a particular risk materializes.

Customers should understand that asking suppliers to accept a greater proportion of risk will have a direct impact on the level of charges payable under the contract. For example, the transfer to the supplier of risk which would be best managed by the customer will lead to an increase in charges. Customers need to take a realistic approach to risk allocation; there are some risks which cannot simply be "outsourced" to suppliers.

Indeed, attempting to do so could damage their core business and potentially put them in breach of their regulatory obligations. In negotiating contracts, customers should focus on those risks which are real and important to their business. Where the risk is not important, they should take a pragmatic and commercial view in negotiations.

Although buyers expect service delivery performance to match what has been contractually agreed on, as the focus shifts increasingly to cost reduction, providers and clients must be able to manage the relationship and the expectations, not just the contract. Over the lifetime of the deal the focus is likely to shift rapidly toward improvement and innovation. The challenge will be in managing change while maintaining realistic expectations among the key deal stakeholders within the context of the deal.

To gain advantage from the major trends driving the rapid evolution of infrastructure outsourcing, buyers must accept — and when possible seek — opportunities for consolidation, industrialization and global delivery. The best outsourcing contracts ensure the proverbial win-win relationship, where both parties arrive at terms and pricing that is fair and promises long-term sustainable value. Contracts that leave the provider with only marginal profit or limited revenue growth will ultimately result in service quality issues.

In setting up these managed services contracts, operators' main goal has been improved cost efficiency — a result of the outsourcing vendor's ability to realize economies of scale, and the global knowledge and experience of the delivery partner. In general, the decision to contract for managed services is made at operating-company level based on local circumstances. While managed services contracts have cost efficiency built in, operators still need to see hard evidence of major cost advantages before they will be convinced to outsource more activities according to Ey.

There is fast growing trend toward sale and leaseback of tower infrastructure certainly in Africa. Telcos use of this technique varies on a regional basis, reflecting a lack of infrastructure players in some markets, and the fact that some regulatory authorities do not allow it. There is also a degree of uncertainty over the extent of the benefits to operators — and over whether the tower companies will share the same objectives as the operators in the future. In tower transactions, the contract and counterparties involved vary from an outright sale to a third-party infrastructure provider, to a transfer of assets to a tower company created and wholly owned by the operator itself, either on a stand-alone basis or in a joint venture.

Selecting the most appropriate supplier for a project can reduce risk substantially. It is incumbent on a customer to devise a due diligence process that will properly test and evaluate potential suppliers. A successful due diligence exercise should not just be a paper exercise; it should involve visiting potential suppliers, testing technology and speaking to other customers of the supplier. It is also important for customers to consider soft issues such as cultural fit. All too often outsourcing goes wrong because it was not possible to create an effective working partnership between customer and supplier.

It is clear that whilst the general economic situation remains difficult, customers will continue to be under pressure to reduce their costs and will look to their suppliers to help them do so. When it comes to entering into new arrangements customers should draft their contracts to allow for the maximum possible flexibility. No business or business environment remains static – change is inevitable. The contract should therefore contain a mechanism for managing contractual and operational change. Once again, good governance is key.

Through creative partnering and innovative risk sharing options, new managed services and outsourcing business model options provide the framework for creating a next generation enabled portfolio of services for consumers and enterprises ready for Telco 2 sided business model. Instead of short-term tactical advantages, the focus is firmly on long-term strategic gains . And only then will Telcos be able to reap the full benefits of managed services via trusted partnerships.

---♠---

5.7 : Location Insight Services (LIS) : Turning BIG data into BIG $$

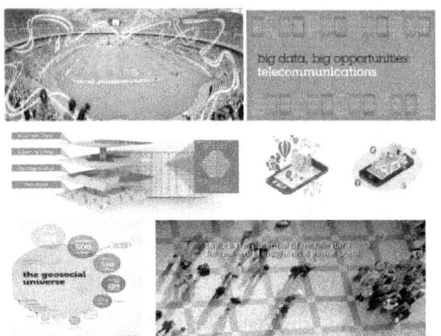

Recent research by (JDSU / STL) has revealed an US$11Bn global opportunity for operators to monetize the data in their networks about places and people. The study concluded that demand for what it calls location insight services (LIS) will be driven predominantly by retailers that want to know more about local market trends and benchmark themselves against their competitors. Telcos are uniquely positioned to capitalise on LIS, as opposed to location-based services (LBS), which is proving more lucrative for over-the-top (OTT) service providers than telcos.

For some time the mobile industry has focused heavily on the opportunity presented by real time Location Based Services (LBS) for individual subscribers, a market that is estimated to reach $12.7 billion by 2014, according to Juniper Research. While there has been great success with LBS for apps targeted at consumers, many mobile operators have struggled to realize their share of this opportunity, with most of the revenue going to over-the-top (OTT) content players. OTT players lead the way in using real-time location data to provide location-centric services to consumers, such as special offers or vouchers.

By contrast, the Location Insight Services segment offers operators a new opportunity to monetize their location data. Telcos have a clear advantage over OTT players because they can aggregate huge volumes of anonymous location data over time and delivering value either directly to businesses, or via partners such as retail planners and advertising agencies. The underlying premise is that identification of repetitive patterns in location activity over time not only enables a much deeper understanding of the consumer in terms of behaviour and motivation, but also builds a clearer picture of the visitor profile of the location itself.

LIS plays to the strengths of operators because their engineers already collect anonymous location data for the purposes of analysing network performance and capacity planning.Various analysts have confirmed that there is a massive latent demand for location-centric information within the business community to enable the delivery of location-specific products and services that are context-relevant to the consumer. According to the Economist Business Unit, there is a consensus amongst marketers that location information is an important element in developing marketing strategy, even for those companies where data on customer and prospect location is not currently collected

LIS is an extension of existing software and analytics systems although data collected by these systems requires additional processing before it can be re-packaged into something marketable.This information can be shared with external systems and can be integrated with data warehouses using cost effective techniques. In many cases the intelligence can be directly used with business intelligence solutions. While commonly available cell level location enables some of the use cases, building level location intelligence from a carrier grade LIS system significantly increases the value. Examples of LIS include:

• Competitive Benchmarking (Retail) – previously unavailable intelligence on the profile of visitors to competitive stores
• Infrastructure Planning (Transport) – clear identification of "pinch points" on transport infrastructure and the precise times they occur

• Site Selection (Event planning) – evaluating previous attendee levels at a venue and attendance at competitive events with a similar audience profile
• Advertising Evaluation (Advertising/Retail) – determining the impact of advertising on store visits

For example , LIS platforms can enable mobile operators to share precise location data with transport infrastructure planners to help the understanding of where in the transport network heavy traffic occurs and when. This insight can be used to plan effective investment in infrastructure, and increase citizen satisfaction by improving transport network efficiency. LIS platforms provide the trend insight about which venues receive the best audience attendances given certain parameters, which can then be used to create a framework for predictive audience modelling. This enables event planners to more accurately assess the viability of venue locations, without needing to carry out time and resource intensive customer research.

Some Tier 1 Telcos have recognized the opportunity and publicly made noises about providing this insight. Last year Telefonica Digital unveiled a new division called Telefonica Dynamic Insights, which is tasked with monetising its vast data resources. Their first product, 'Smart Steps', will use fully anonymised and aggregated mobile network data to enable companies and public sector organisations to measure, compare, and understand what factors influence the number of people visiting a location at any time.

These insights will help retailers tailor local offerings for existing stores and determine the best locations and most appropriate formats for new stores. A number of retailers are already helping with product development by providing user feedback. Smart Steps will also be able to help town councils measure how many more people visit their high street after the introduction of free car parking, farmers markets, or late night shopping.

Big data is one of the more fascinating developments in today's tech world: harnessing the huge wave of information that comes out of Internet-based networks and then trying to make sense of it. Mobile operators have huge repositories of data in their businesses : not just from people's activity on cellular networks, but from WiFi networks, too. LIS puts the power back in operators' hands allowing them to monetise the value of their unique asset, mass location intelligence, creating new revenue streams in times where traditional business models remain under extreme pressure. Hey guys, it is time to stop complaining about OTT marauders and take action to monetise one of your network's biggest yet most untapped asset : Location Location Location !!

--♠--

5.8 : Telco Growth Imperative : How to organise for new services

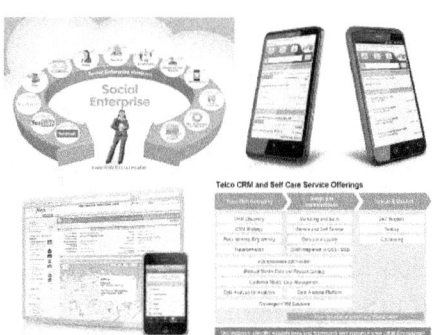

Telecommunications companies are facing difficult times these days: their products are commoditised, competition is fierce and pressure on margins is high. Some operators with foresight are trying to set themselves apart through better service. Services, by being less visible and more labor dependent, are much more difficult to imitate, thus becoming a sustainable source of competitive advantage.

Reaching out to and understanding the needs of customers, both current ones and those who may consider shifting from competitors in such turbulent and competitive era, is an important element of the service strategy. As such Telcos would do well to learn from other industry best practices to gain a competitive edge in and expand beyond their core markets.

Other industries, such as computer or telecom hardware, have shown that business development based on new services can be a successful road to new growth. In 1993, CEO Louis Gerstner initiated a transformation at IBM. The transformation involved a change from a hardware and software business to a solutions and services business and from a regionally aligned organization to a global organization. IBM committed itself to business and cultural change , invested in talent and the right financial and IT systems to support them and placed strategic bets on IT as a utility service and hosted storage. The transformation created opportunities for cost savings by encouraging development and use of enterprise-wide technology platforms.

Fortunately with the ever evolving technology and more complex products, incumbent operators have a powerful asset to leverage: their technical field service organization and capabilities. In order to capitalise on further strategic growth opportunities, Telcos should consider an option of developing a dedicated service organization which has control of its entire value chain, primarily focusing on multiproduct communications services for mass segments. This means a dedicated unit, focusing on developing the service business, having full control over its entire value chain, freedom and leeway to develop its business, full management attention and support to execute its mandate.

The primary focus of the new service organizations is service innovation excellence and ability to scale customer solutions for rapid growth across well defined customer segments based on their real communications needs. Customer front-end responsibility (marketing and sales) should be placed into new service Org. The consolidation of the service offering under a single division is normally accompanied by a strong initiative to improve the efficiency, quality and delivery time of the services provided, and the creation of additional services to supplement the basic offering. The consolidation of services also comes with the development of a monitoring system to assess the effectiveness and efficiency of the service delivery.

Transitioning from product manufacturer into service provider constitutes some managerial challenges. Services require new organizational principles, structures and processes. Not only are new capabilities, metrics and incentives needed, but also the emphasis of the business model changes from transaction- to relationship-based. Be warned that research has shown it takes a serious effort by senior managment to build the structures, capabilities, processes and systems to seize the service opportunities.Successful service companies do not start from scratch – they are built up on the basis of existing units and businesses with the best suited set of service assets and capabilities such as customer knowledge ; service development, standardization and roll-out capabilities for complete service delivery.

A focus on forward-looking IT investments (funded by reductions in maintenance costs for today's systems) will be essential to support the service organisation. Social media collaboration platforms support all service operations: enterprise case management, call centers, customer portals, websites, and integration with social media channels. A knowledge base provides answers to your agents and your customers through all your channels, increasing deflection rates and reducing time spent per case, keeping your customers happy and loyal. Cloud based CRM platforms can support the customer service team to improve the way they managed everyday customer interactions.

Communications service providers have a number of attributes that give them a potential marketplace advantage: an extensive customer base, distribution muscle and knowledge of customer preferences through CRM and billing systems. The opportunity is to become an integrated digital services provider across platforms and mobile devices—convincing customers that a communications service provider can effectively serve as the hub to meet their communication and entertainment needs. According to an Accenture survey, the areas that show particular promise include cloud services and location-based offers.

Cloud Customer Portals give customers a true online service experience, ensuring they have the flexibility to manage their interactions with their Telco entirely online if they chose to do so, and enabling customer service questions to be managed, just like an order coming in from a field sales team person. Customers can create orders online for new and replacement products including, phones, accessories, and SIM cards, and then track the status of the order through to shipment. When Sprint acquired Nextel, over 5,000 employees in over 1,100 retail stores and 800 dealer locations got busy collaborating on customer retention and churn avoidance. Tied together via an employee social network, disparate teams across both organizations focused their efforts to retain customers and build new loyalty programs.

Delivering good services as part of a core product offering does not suffice as the sole differentiator in highly competitive telecommunications markets. Investments in new radio access technology bring along radically new network economics leaving mobile operators with the quest to gear their network investments towards a cost optimal access, backhaul and core portfolio. It is critical to cut spending on low-value activities, and redeploy it to investments that generate growth, margins and true differentiation. Being able to accurately identify where value is generated at all levels of the organization – from divisions to specific products or offerings to particular customers – is a critical managerial competence.

Customer ownership and distribution power give communications service providers a strong foundation on which to build to meet consumers' ongoing communication and entertainment needs. Providers have an opportunity to improve their return on investment by monetizing better connectivity. They can also extend their partnerships across the digital ecosystem to provide a seamless customer experience. This will require deep insight into subscriber behaviors, new forms of collaboration within the industry, new capabilities within the organization and an ability to constantly innovate to keep pace with today's demanding consumers.

Perhaps one of the most successful new age " experience " players to date is SK Planet, which was set up in 2011 by SK Telecom, Korea's largest wireless operator, to offer multiple add-on experiences for both retail and business subscribers. They include MelOn, already Korea's largest music portal, with 17 million subscribers, has also been launched in Indonesia; 11st provides an e-commerce platform with related advertising and marketing intelligence services.It is now the country's second-largest e-commerce platform and largest player in mobile commerce; "T ad" is a mobile ad platform that enables personalized ads on mobile apps running on smartphones and tablets; "T map," a GPS-based navigation service platform with more than 10 million subscribers, also offers location-based services to businesses.

By consolidating a wide range of services under one roof, on top of its successful core wireless broadband business, SK Planet now offers perhaps one of the most compelling customer experiences of any operator worldwide.To provide customers with exactly the right products and services based on their actions and requests, " experience " players such as SK have to become fully responsive to the correct interpretations of their customers' behavior, often in real time. The ability, for instance, to offer access to medical information services could follow evidence of increased interest in healthcare. So investing in data analytics capabilities to respond to customer data is a must. "Big data" offers much promise in this area, but it will require considerable investment.

According to experts at Booz " experience play " Telcos de-emphasize their network activities and in some cases even carving out their entire access network infrastructure—both passive and active. They can share these costly assets with competitors via network-sharing agreements, and then differentiate themselves through innovative products and services.

 In contrast, some recent mobile and fixed entrants in Europe and the Middle East have focused on deploying their own network infrastructure in order to become connectivity or platform players. Consequently, they have minimized their investments in customer facing infrastructure, relying on an online presence for sales and deploying only flagship stores to serve as their bricks-and-mortar channel.

Telecom operators should look for opportunities for growth by both assessing their markets (the market-back view) and evaluating their own current strengths (the capabilities-forward view). The capabilities forward view seeks to find distinctive internal capabilities that can be leveraged in any number of ways: to grow into adjacent markets, to build innovative new services, or to increase network speed and capacity. The market-back view turns outward for market opportunities that might arise from new technologies or from opportunities that competitors might be overlooking or are not coherently pursuing.

The goal is to become coherent: to strike a balance so that the right product and service portfolio naturally thrives within a capabilities system consciously chosen implemented.

-------------------------------------THE END-------------------------------------

www.ingramcontent.com/pod-product-compliance
Lightning Source LLC
Chambersburg PA
CBHW080808180526

45168CB00006B/2363